AF694886

FINITE ELEMENT ANALYSIS OF THIN-WALLED STRUCTURES

FINITE ELEMENT ANALYSIS OF THIN-WALLED STRUCTURES

Edited by

JOHN W. BULL

Department of Civil Engineering,
University of Newcastle upon Tyne, UK

CRC Press
Taylor & Francis Group
Boca Raton London New York

CRC Press is an imprint of the
Taylor & Francis Group, an **informa** business
A TAYLOR & FRANCIS BOOK

CRC Press
Taylor & Francis Group
6000 Broken Sound Parkway NW, Suite 300
Boca Raton, FL 33487-2742

First issued in paperback 2019

ISBN-13: 978-1-85166-136-7 (hbk)
ISBN-13: 978-0-367-86582-5 (pbk)

British Library Cataloguing in Publication Data

Finite element analysis of thin-walled
structures
1. Shells (Engineering) 2. Finite element
method
I. Bull, J. W.
624.1'7762 TA660.S5

Library of Congress Cataloging-in-Publication Data

Finite element analysis of thin-walled structures/edited by John W. Bull.
p. cm.
Bibliography: p.
Includes index.
ISBN 1-85166-136-0
1. Thin-walled structures. 2. Finite element method.
3. Structures, Theory of--Data processing. I. Bull, John W.
TA660.T5F56 1988
624.1'71--dc19 87-20005 CIP

Visit the Taylor & Francis Web site at
http://www.taylorandfrancis.com

and the CRC Press Web site at
http://www.crcpress.com

Preface

The aim of this book is to present a series of chapters that describe current developments in the finite element analysis and design of certain types of thin-walled structures. Each chapter is presented in a self-contained manner and concentrates on the practical aspects of finite elements and their use.

The choice of using the finite element method of analysis to the exclusion of other methods of analysis is, for the user, considerably eased by the ready availability of finite element analysis packages, coupled with the easy access to the required computing resources.

For engineers, an answer to within 10% of the correct answer is acceptable. To a mathematician, 1% is a large error, but mathematicians usually have alternative analytical solutions with which to calibrate their results. No such analytical calibration is available to an engineer who may be designing an air-inflated building, an axisymmetric thin shell, a ship's hull or the leg of an oil-drilling platform.

In the analysis of shell structures using finite elements, which the editor researched for his Ph.D., care had to be taken to ensure that elements were always used within the boundaries of their theoretical development. This proved to be no easy task. Elements that gave accurate answers to, say, a shallow barrel-vault roof could be most inaccurate when used for a deep barrel-vault roof.

In design offices, engineers are under pressure, and rightly so, to produce safe and efficient designs. Engineers need to have a knowledge of the available finite elements and their associated software, together with an ability to detect and avoid incorrect numerical results. The first three chapters of this book are designed to assist in this process.

Chapter 1 lays the foundations for the development and the use of finite elements for thin-walled structures. Chapter 2 looks at the availability of finite element software packages and suggests caution with respect to their indiscriminate use. In Chapter 3 the necessity of validating the input data, the elements and the mesh arrangement to be used are discussed. The chapter even suggests the testing of the elements to be used.

In deciding which area of structures to apply the use of finite elements to, a library search revealed that, although there are over 50 books on finite elements and 13 or so books on thin-walled structures published in the UK, there was no one book bringing together both finite elements and thin-walled structures. This was surprising, considering the rapidly increasing use of structures, such as buildings, pipelines, ships and oil platforms that rely upon their in-plane stiffness to resist loading. The final four chapters of this book show how the finite element method is used to assist in the solution of these thin-walled structures.

Chapter 4 analyses membrane structures varying from cables under self weight to water-inflated dams and air-inflated buildings. In Chapter 5 an axisymmetric thin-shell element has been developed and applied to a series of problems. Chapter 6 shows in detail the relationship between the finite element method, the loads, stresses, analytical and experimental work on ship structures and how this relationship is changing the way in which ships are designed. Finally, Chapter 7 illustrates details of offshore structures that are most frequently analysed using thin-shell elements.

John W. Bull

Contents

List of Contributors

ED AKIN

Department of Mechanical Engineering and Materials Science, Rice University, 6100 Main Street, P.O. Box 1892, Houston, Texas 77251, USA

DAVID ANDREWS

Sea System Controllerate, Ministry of Defence, Foxhill, Bath BA15AB, UK.

SLADE GELLIN

Bell Aerospace Textron, P.O. Box 1, Buffalo, New York 14240, USA

D. HITCHINGS

Department of Aeronautics, Imperial College of Science and Technology, Prince Consort Road, London SW7 2AZ, UK

NAHID JAVADI

Wimpey Offshore, 27 Hammersmith Grove, Hammersmith, London W6 7EN, UK. Address for correspondence: *56 Bolingbroke Rd, London W14, UK*

GEORGE C. LEE

Faculty of Engineering and Applied Science, State University of

New York, at Buffalo, 412 Bonner Hall, Buffalo, New York 14260, USA

C. T. F. Ross

Department of Mechanical Engineering, Portsmouth Polytechnic, Anglesea Building, Anglesea Road, Portsmouth PO1 3DJ, UK

C. H. Woodford

Computing Laboratory, University of Newcastle upon Tyne, Claremont Road, Newcastle upon Tyne NE1 7RU, UK

Chapter 1

Finite Elements Available for the Analysis of Non-curved Thin-walled Structures

SLADE GELLIN
Bell Aerospace Textron, Buffalo, New York, USA

and

GEORGE C. LEE
State University of New York at Buffalo, New York, USA

1.1. INTRODUCTION

The purpose of this chapter is to give a theoretical basis for some sample finite elements which can be used to analyse long, slender straight members whose cross-sections can be classified as thin-walled. This theoretical basis can be extended to develop finite elements appropriate to curved members, or more complex behavior to be experienced by the straight members.

While most readers of this book know what the finite element method is, it may be useful to identify its place in the set of formulation schemes available for structural mechanics. Generally, there are three alternate ways of posing a structural mechanics problem. These ways are (i) as differential equations, (ii) energy methods and (iii) integral equations. The second and third of these are usually derived from the first, most familiar method. Associated with each of these approaches to structural mechanics problems are *approximate* procedures which tend to discretize all or part of the structure in the sense that the behavior of a discrete number of points in the structure characterizes the overall response of the structure to loading. Finite difference methods can be thought of as associated with the differential equations approach. Finite element methods are extensions of the energy methods. The boundary element method is used with an integral equation formulation.[1] It is the second of these that concerns us here; thus, an understanding of the energy methods applicable to non-curved thin-walled structures is necessary in order to eventually formulate finite elements for these members.

Generally, energy methods derive from the application, extension and sometimes combination of two basic principles. The first of these is the principle of virtual work. Suppose a body is in equilibrium. If it is imagined that an admissible displacement field $\delta\mathbf{u}$ is added to the current displacements, then the principle states that the work done by the stress field $\boldsymbol{\sigma}$ through the virtual strains $\delta\boldsymbol{\varepsilon}$ (formulated by using $\delta\mathbf{u}$ in the strain–displacement laws) equals the work done by external forces $\mathbf{F}$ through the virtual displacements $\delta\mathbf{u}$. Mathematically, this can be written as

$$\boldsymbol{\sigma} * \delta\boldsymbol{\varepsilon} = \mathbf{F} * \delta\mathbf{u} \tag{1.1}$$

where the operator $*$ is appropriate for the structure at hand. The left-hand side of eqn (1.1) is referred to as the internal virtual work, while the right-hand side is called the external virtual work. In most applications a constitutive law exists relating $\boldsymbol{\sigma}$ with $\boldsymbol{\varepsilon}$, and, by using calculus of variations, eqn (1.1) can be manipulated to obtain equations for the displacement field $\mathbf{u}$.

The term 'admissible' for the virtual displacement $\delta\mathbf{u}$ implies that two conditions are satisfied by $\delta\mathbf{u}$. The first of these is that $\delta\mathbf{u}$ does not violate any existing displacement boundary conditions; and secondly, that $\delta\mathbf{u}$ is continuous and differentiable enough times in order to produce a non-trivial strain field. For example, suppose that the strains are first-order derivatives of the displacements; then, the displacement field must be continuous so that the first derivatives may be taken. These first derivatives, however, need *not* be continuous.

The solution of eqn (1.1) for $\mathbf{u}$ may be very difficult if the geometry is difficult to describe for a particular co-ordinate system, or if the equation itself (usually a partial differential equation) does not have simple general solutions. The best recourse at this point is to guess a particular form of $\mathbf{u}$ with undetermined parameters that is admissible, and, with use of the strain–displacement and constitutive laws, use in eqn (1.1) directly, deriving equations for the undetermined parameters. It should be realized that, in general, only by guessing the exact solution will the resulting stress field be in equilibrium with the applied loads and satisfy the stress boundary conditions. All other solutions produced from other guesses, while not satisfying equilibrium, can at least be considered the 'best' that a particular choice of $\mathbf{u}$ can give, if we consider satisfaction of eqn (1.1) as a criterion.

The other basic principle used in energy methods is the principle of complementary virtual work. Suppose a body has a compatible strain field. If it is imagined that an admissible stress field $\delta\boldsymbol{\sigma}$ is added to the current stresses, then the principle states that the complementary work done by the strain field $\boldsymbol{\varepsilon}$ through the virtual stress $\delta\boldsymbol{\sigma}$ equals the complementary work done by the displacement field through the virtual forces. Mathematically, this can be written as

$$\boldsymbol{\varepsilon} * \delta\boldsymbol{\sigma} = \mathbf{u} * \delta\mathbf{F} \tag{1.2}$$

In most applications the constitutive laws are used in the left-hand side of eqn (1.2) and, by using the calculus of variations, eqn (1.2) can be manipulated to yield resulting compatibility equations for $\boldsymbol{\sigma}$. Here, an admissible stress field is one that satisfies the equilibrium equations and traction boundary conditions. Equation (1.2) may be used directly to obtain approximate results. Choosing an admissible stress field with undetermined parameters will yield the 'best' strain field for that stress field choice; however, only the exact solution will yield a compatible strain field.

The finite element method, then, is a procedure which chooses admissible fields over finite regions of the structure, such that when assembled, the field is admissible over the entire structure, and then uses eqns (1.1), (1.2) or derived combinations to obtain approximate solutions.

Section 1.2 will be devoted to devising appropriate forms of eqns (1.1) or (1.2) for thin-walled structures. In subsequent sections, simple finite elements for linearly elastic members will be derived, and two advanced extensions of the simple element will be examined. Numerical examples will be given, and the chapter will be summarized.

1.2. VARIATIONAL PRINCIPLES FOR THIN-WALLED STRUCTURES

Figure 1.1 represents a generalized open, thin-walled cross section for an assumed straight structural member of length L. The thickness t is assumed small when compared to the length m of the centerline; this is

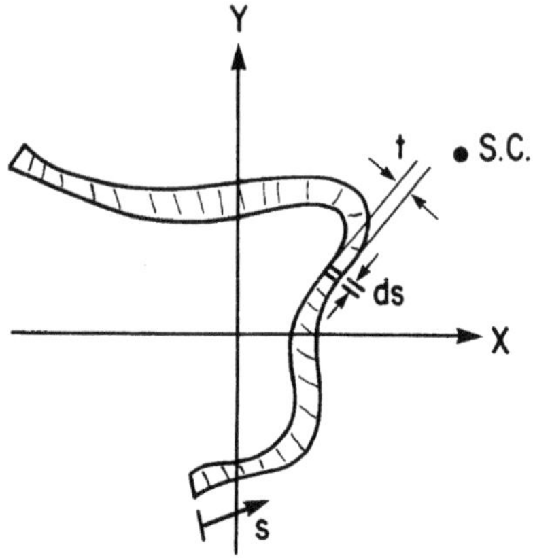

FIG. 1.1. Open, thin-walled cross-section.

basic to the classification as a thin-walled structure. The member can be subject to a variety of loads which result in a stress distribution on the cross-section. The stresses are normal to the surface (denoted as σ) and shear, generally in a direction tangent to the midline of the thin section (denoted as τ). Generally, for elastic behavior the following loads give rise to the following stresses:

Axial thrust: → normal

Bending
- moments: → normal
- shears: → shear

Torsion
- St. Venant → shear
- warping → shear, normal

In simple strength of materials theory, there are associated normal strains with the normal stresses and shear strains with the shear stresses; however, the bending shear strains and the warping shear strains are generally ignored compared with their normal counterparts. We will begin with this assumption, and examine the individual load cases for strain–displacement relations.

1.2.1. Axial Thrust

The simplest case, axial thrust, results in an axial displacement w as a function of z only; thus, the only strain is

$$\varepsilon^{\text{thrust}} = \mathrm{d}w/\mathrm{d}z \tag{1.3}$$

1.2.2. Bending

For pure bending the axial displacement varies linearly with the distance from the bending axis, the constant of proportionality being the rotation about that axis. In terms of u and v, the displacement in the x and y directions of the centerline, these rotations are

$$\phi_x = -\mathrm{d}v/\mathrm{d}z; \qquad \phi_y = \mathrm{d}u/\mathrm{d}z \tag{1.4}$$

Thus the total axial displacement is

$$W = y\phi_x - x\phi_y \tag{1.5}$$

Differentiating yields

$$\varepsilon^{\text{bending}} = y\frac{\mathrm{d}\phi_x}{\mathrm{d}z} - x\frac{\mathrm{d}\phi_y}{\mathrm{d}z} \tag{1.6}$$

1.2.3. Torsion

The torsion displacement quantity is the angle $\phi_z(z)$ that the cross-section at z rotates about a z-axis running through the shear center, which in general, would not coincide with the centroid of the cross-section. The cross-section will also warp out of its undeformed plane. When the warping is unrestricted, only the so-called St. Venant stresses and strains are present. Generally, for thin-walled sections, the St. Venant torsion effects are assumed to be those of a rectangle, one dimension of which becomes much smaller than the other. Curved sections can be thought of as connected infinitesimal thin straight sections. As $t/m \to 0$, then

$$\gamma_{zs} \approx 2n\frac{\mathrm{d}\phi_z}{\mathrm{d}z}; \qquad \gamma_{zn} = 0 \tag{1.7}$$

away from the endpoints of the section, $s = 0$ and $s = m$, where n is the normal distance from the centerline. Near the ends, τ_{zs} approaches zero while τ_{zn} takes on a finite value. Generally, the ends can be ignored as the transition is rather rapid near those ends only; furthermore, though it can be shown that the end τ_{zn} contributes to the twisting moment, the associated strain energy is quite small; thus, since γ_{zs} is the γ of interest, we rewrite eqn (1.7) as

$$\gamma^{sv} = 2n\frac{\mathrm{d}\phi_z}{\mathrm{d}z} \tag{1.8}$$

The warping normal displacement is usually found by assuming that the angle of inclination of the cross-section at a particular point on the centerline is proportional to the rate of twist $d\phi_z/dz$ and the perpendicular distance from the shear center to the tangent line at the point on the midline in question. With this distance denoted as r, this can be mathematically represented as

$$\frac{dW}{ds} = -r\frac{d\phi_z}{dz} \tag{1.9}$$

Integrating eqn (1.9) from 0 to s yields

$$W(s) = W(0) - \omega_s \frac{d\phi_z}{dz} \tag{1.10}$$

where

$$\omega_s = \int_0^s r\,ds \tag{1.11}$$

If no axial thrust is present, the average value of W should be zero over the cross-section. Integrating eqn (1.10) from 0 to m and dividing by m thus yields

$$W(0) = \bar{\omega}_s\, d\phi/dz \tag{1.12}$$

where

$$\bar{\omega}_s = \frac{1}{m}\int_0^m \omega_s\, ds \tag{1.13}$$

Substituting back into eqn (1.10) then gives

$$W = (\bar{\omega}_s - \omega_s)\frac{d\phi_z}{dz} \tag{1.14}$$

When warping is resisted, $d\phi_z/dz$ is not a constant, but varies with z; thus, the normal strain associated with warping is

$$\varepsilon^{\text{warping}} = (\bar{\omega}_s - \omega_s)\frac{d^2\phi_z}{dz^2} \tag{1.15}$$

Combining eqns (1.3), (1.6), (1.8) and (1.15) gives

$$\varepsilon = \frac{dw}{dz} - y\frac{d^2y}{dz^2} - x\frac{d^2u}{dz^2} + (\bar{\omega}_s - \omega_s)\frac{d^2\phi_z}{dz^2} \tag{1.16a}$$

$$\gamma = 2n\frac{d\phi_z}{dz} \tag{1.16b}$$

We now use eqn (1.16) in an appropriate form for the internal virtual work. This form is

$$\delta(IVW) = \int_v (\sigma\delta\varepsilon + \tau\delta\gamma)\,\mathrm{d}V$$
$$= \int^L \int_0^m \int_{-t/2}^{t/2} \sigma\delta\left[\frac{\mathrm{d}w}{\mathrm{d}z} + y\frac{\mathrm{d}\phi_x}{\mathrm{d}z} - x\frac{\mathrm{d}\phi_y}{\mathrm{d}z} + (\bar{\omega}_s - \omega_s)\frac{\mathrm{d}^2\phi}{\mathrm{d}z^2}\right]$$
$$+ \tau\delta\left[2n\frac{\mathrm{d}\phi_z}{\mathrm{d}z}\right]\mathrm{d}n\,\mathrm{d}S\,\mathrm{d}z \qquad (1.17)$$

The following stress resultants may be identified:

$$\left.\begin{aligned}
P &= \int_0^m \sigma t\,\mathrm{d}s && \text{(axial thrust)}\\
M_x &= \int_0^m \sigma y t\,\mathrm{d}s && \text{(bending moment about } x\text{-axis)}\\
M_y &= -\int_0^m \sigma x t\,\mathrm{d}s && \text{(bending moment about } y\text{-axis)}\\
M_\omega &= \int_0^m \sigma(\bar{\omega}_s - \omega_s) t\,\mathrm{d}s && \text{(warping bending moment)}\\
T_{sv} &= \int_0^m \int_{-t/2}^{t/2} 2n\tau\,\mathrm{d}n\,\mathrm{d}s && \text{(St. Venant torque)}
\end{aligned}\right\} \qquad (1.18)$$

Substituting eqn (1.18) into eqn (1.17) yields the general internal virtual work expression for an open thin-walled section of length L:

$$\delta(IVW) = \int_0^L \left(P\,\delta\frac{\mathrm{d}w}{\mathrm{d}z} + M_x\delta\frac{\mathrm{d}\phi_x}{\mathrm{d}z} + M_y\delta\frac{\mathrm{d}\phi_y}{\mathrm{d}z} + M_\omega\delta\frac{\mathrm{d}^2\phi_2}{\mathrm{d}z^2} + T_{sv}\delta\frac{\mathrm{d}\phi_z}{\mathrm{d}z}\right)d \qquad (1.19)$$

Note that no constitutive law has been applied in deriving eqn (1.19).

The external virtual work is calculated by examining the individual loads that this structural member can support. Generally, it will support point forces and moments at its ends, and distributed forces and moments per unit length. This can be represented as

$$\delta(EVW) = \int_0^L (q_x\delta u + q_y\delta v + q_z\delta w + m_x\delta\phi_x + m_y\delta\phi_y + m_z\delta\phi_z)\,\mathrm{d}z$$
$$+ F_x^0\delta u(0) + F_x^L\delta u(L) + F_y^0\delta v(0) + F_y^L\delta v(L)$$
$$+ F_z^0\delta w(0) + F_z^L\delta w(L)$$
$$+ C_x^0\delta\phi_x(0) + C_x^L\delta\phi_x(L) + C_y^0\delta\phi_y(0)$$
$$+ C_y^L\delta\phi_y(L) + C_z^0\delta\phi_z(0) + C_z^L\delta\phi_z(L) \qquad (1.20)$$

Expressions (1.19) and (1.20) are equal by eqn (1.1). Equilibrium equations and appropriate boundary conditions could be derived by integration by parts and calculus of variations. In addition, eqn (1.4) may be substituted into eqn (1.19) and/or eqn (1.20) at various stages, either as constraints on two 'independent' variables, or directly.

If the material behavior is characterized by linear elasticity, then

$$\sigma = E\varepsilon; \qquad \tau = G\gamma \tag{1.21}$$

Substituting eqn (1.21) into the first line of eqn (1.17) yields

$$\delta(IVW) = \int_v (E\varepsilon\delta\varepsilon + G\gamma\delta\gamma)\,\mathrm{d}V \tag{1.22}$$

But since

$$E\varepsilon\delta\varepsilon = \delta(\tfrac{1}{2}E\varepsilon^2) \tag{1.23}$$

then

$$\delta(IVW) = \delta\left(\int_v (\tfrac{1}{2}E\varepsilon^2 + \tfrac{1}{2}G\gamma^2)\,\mathrm{d}V\right) \tag{1.24}$$

Expression (1.20) is in the form $\mathbf{F} * \delta\mathbf{u}$. Since the F are given, then

$$\delta(EVW) = \delta(\mathbf{F} * \mathbf{u}) \tag{1.25}$$

Subtracting eqn (1.25) from eqn (1.24) yields

$$\delta(IVW - EVW) = \delta\pi_{\mathrm{p}} = \delta\left[\int_v (\tfrac{1}{2}E\varepsilon^2 + \tfrac{1}{2}G\gamma^2)\,\mathrm{d}V - F * u\right] = 0 \tag{1.26}$$

The functional π_{p} is called the potential energy functional for the structure. It is a special case of the principle of virtual work. What eqn (1.26) says is that of all admissible displacement fields, the ones satisfying equilibrium render a stationary value for π_{p}; and it can be shown that, for stable equilibrium, π_{p} is a minimum. These statements make up what is known as the minimum potential energy principle, which is probably the most applied in elementary finite element analysis. What is most significant for our purposes here is to realize that for a given subset of admissible displacement fields, the member of the subset that gives the lowest value possible for π_{p} will be considered the best approximation to the actual solution that that subset can supply.

Substituting eqn (1.16) into eqn (1.26) yields

$$
\begin{aligned}
\pi_{\mathrm{p}} = \tfrac{1}{2}\int_0^L \Bigg(E\Bigg[& \left(\int_0^m t\,\mathrm{d}s\right)\left(\frac{\mathrm{d}w}{\mathrm{d}z}\right)^2 + \left(\int_0^m y^2 t\,\mathrm{d}s\right)\left(\frac{\mathrm{d}\phi_x}{\mathrm{d}z}\right)^2 \\
& + \left(\int_0^m x^2 t\,\mathrm{d}s\right)\left(\frac{\mathrm{d}\phi_y}{\mathrm{d}z}\right)^2 + \left(\int_0^m (\bar{\omega}_s - \omega_s)^2 t\,\mathrm{d}s\right)\left(\frac{\mathrm{d}^2\phi_z}{\mathrm{d}z^2}\right)^2 \\
& + 2\left(\int_0^m yt\,\mathrm{d}s\right)\frac{\mathrm{d}w}{\mathrm{d}z}\frac{\mathrm{d}\phi_x}{\mathrm{d}z} - 2\left(\int_0^m xt\,\mathrm{d}s\right)\frac{\mathrm{d}w}{\mathrm{d}z}\frac{\mathrm{d}\phi_y}{\mathrm{d}z} \\
& - 2\left(\int_0^m xyt\,\mathrm{d}s\right)\frac{\mathrm{d}\phi_x}{\mathrm{d}z}\frac{\mathrm{d}\phi_y}{\mathrm{d}z} + 2\left(\int_0^m (\bar{\omega}_s - \omega_s)t\,\mathrm{d}s\right)\frac{\mathrm{d}w}{\mathrm{d}z}\frac{\mathrm{d}^2\phi_z}{\mathrm{d}z^2} \\
& + 2\left(\int_0^m y(\bar{\omega}_s - \omega_s)t\,\mathrm{d}s\right)\frac{\mathrm{d}\phi_x}{\mathrm{d}z}\frac{\mathrm{d}^2\phi_z}{\mathrm{d}z^2} \\
& + 2\left(\int_0^m x(\bar{\omega}_s - \omega_s)t\,\mathrm{d}s\right)\frac{\mathrm{d}\phi_y}{\mathrm{d}z}\frac{\mathrm{d}^2\phi_z}{\mathrm{d}z^2}\Bigg] \\
& + G\left(\int_0^m \tfrac{1}{3}t^3\,\mathrm{d}s\right)\left(\frac{\mathrm{d}\phi_z}{\mathrm{d}z}\right)^2\Bigg)\,\mathrm{d}z - F * u
\end{aligned}
\tag{1.27}
$$

The quantities in parentheses are various geometric properties of the cross-section. In particular:

$$
\left.
\begin{array}{lll}
A = \displaystyle\int_0^m t\,\mathrm{d}s & & \text{(area)} \\[2ex]
I_{xx} = \displaystyle\int_0^m y^2 t\,\mathrm{d}s & & \text{(second moment of area about } x\text{-axis)} \\[2ex]
I_{yy} = \displaystyle\int_0^m x^2 t\,\mathrm{d}s & & \text{(second moment of area about } y\text{-axis)} \\[2ex]
I_{xy} = \displaystyle\int_0^m xyt\,\mathrm{d}s & & \text{(product of area for } x\text{- and } y\text{-axes)} \\[2ex]
J = \displaystyle\int_0^m \tfrac{1}{3}t^3\,\mathrm{d}s & & \text{(St. Venant torsional constant)} \\[2ex]
\Gamma = \displaystyle\int_0^m (\bar{\omega}_s - \omega_s)^2 t\,\mathrm{d}s & & \text{(warping constant)}
\end{array}
\right\}
\tag{1.28}
$$

The other terms are zero for the following reasons:

$\int_0^m yt\,\mathrm{d}s$	1st moment of area about centroidal x-axis is zero
$\int_0^m xt\,\mathrm{d}s$	1st moment of area about centroidal y-axis is zero
$\int_0^m (\bar{\omega}_s - \omega_s)t\,\mathrm{d}s$	warping displacements produce no net thrust
$\int_0^m (\bar{\omega}_s - \omega_s)ty\,\mathrm{d}s$	warping displacements produce no net moment about x-xis
$\int_0^m (\bar{\omega}_s - \omega_s)tx\,\mathrm{d}s$	warping displacements produce no net moment about y-axis

As a result, eqn (1.27) reduces to

$$\pi_\mathrm{p} = \tfrac{1}{2}\int_0^L \left[EA\left(\frac{\mathrm{d}w}{\mathrm{d}z}\right)^2 + EI_{xx}\left(\frac{\mathrm{d}\phi_x}{\mathrm{d}z}\right)^2 + EI_{yy}\left(\frac{\mathrm{d}\phi_y}{\mathrm{d}z}\right)^2 - 2EI_{xy}\frac{\mathrm{d}\phi_x}{\mathrm{d}z}\frac{\mathrm{d}\phi_y}{\mathrm{d}z} + E\Gamma\left(\frac{\mathrm{d}^2\phi_z}{\mathrm{d}z^2}\right)^2 + GJ\left(\frac{\mathrm{d}\phi_z}{\mathrm{d}z}\right)^2\right]\mathrm{d}z - F * u \quad (1.29)$$

We examine eqn (1.29) to determine admissibility conditions. Note that for w, ϕ_x and ϕ_y, only the first derivatives appear in expression (1.29). This implies that only w, ϕ_x and ϕ_y are continuous, with no such restriction on its derivative. Note however, that if eqn (1.4) is substituted into eqn (1.29), then second derivatives of $u, v,$ and already in eqn (1.29), ϕ_z are present. Thus, these three displacements *and* their first derivatives must be continuous. The boundary condition that could be derived would indicate two independent conditions for w to satisfy (one at each end of the member) while for u, v and ϕ_z there would be four independent conditions (two at each end). This information will be used later.

The complementary virtual work approach can be shown to yield

$$\delta(ICVW) = \int_0^L \left\{\frac{\mathrm{d}w}{\mathrm{d}z}\delta P + \frac{\mathrm{d}\phi_x}{\mathrm{d}z}\delta M_x + \frac{\mathrm{d}\phi_y}{\mathrm{d}z}\delta M_y + \frac{\mathrm{d}^2\phi_z}{\mathrm{d}z^2}\delta M_\omega + \frac{\mathrm{d}\phi_z}{\mathrm{d}z}\delta T_{sv}\right\}\mathrm{d}z \quad (1.30)$$

For linearly elastic systems a principle of minimum complementary energy can be derived. The form of the complementary energy functional is

$$\pi_c = \tfrac{1}{2}\int_0^L \left[\frac{P^2}{EA} + \frac{I_{yy}M_x^2 + 2I_{xy}M_xM_y + I_{xx}M_y^2}{E(I_{xx}I_{yy} - I_{xy}^2)} + \frac{M_\omega^2}{E\Gamma} + \frac{T_{sv}^2}{GJ}\right] \mathrm{d}z - u * F \tag{1.31}$$

For one-dimensional structural members, the complementary energy is generally as easy to work with as the potential energy. Note that the variations in eqn (1.31) are to be taken with respect to force parameters. If F in eqn (1.31) is known, then its variation is zero. Its effect is incorporated into the stress resultant measures.

The complementary energy principle is also very useful as a tool to incorporate shear stresses associated with bending effects. As mentioned above, the bending shear stresses and the warping shear stresses do not have strains associated with them in the typical strength of materials approaches; therefore, a virtual work or potential energy approach does not appear to be forthcoming in a straightforward manner. Since these shear stresses are derived from equilibrium considerations, and since the virtual stress field must satisfy equilibrium, it would then appear that a complementary energy approach would be more useful.

The free body diagram shown in Fig. 1.2 indicates that

$$\tau t = -\int_0^s \frac{\mathrm{d}\sigma}{\mathrm{d}z} t \,\mathrm{d}s \tag{1.32}$$

Assuming principal axes, linear elasticity, and both axial bending and

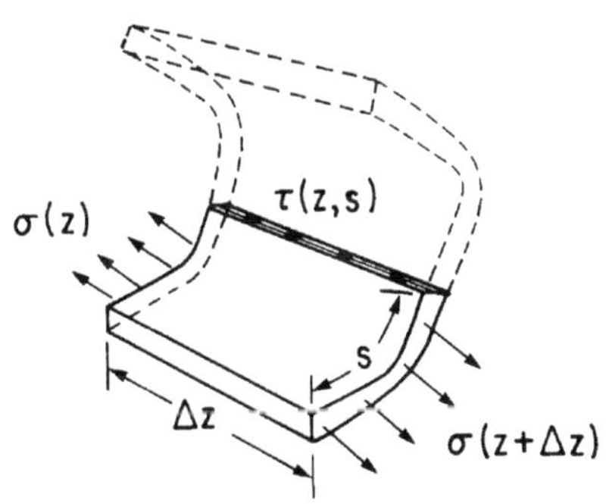

FIG. 1.2. Free body diagram to determine shear stress distributions.

torsion to be present,

$$\sigma = \frac{M_x y}{I_{xx}} - \frac{M_y x}{I_{yy}} + \frac{M_\omega(\bar{\omega}_s - \omega_s)}{\Gamma} \tag{1.33}$$

The shear stress resultants are defined as

$$V_y = \frac{\mathrm{d}M_x}{\mathrm{d}z}; \qquad V_x = -\frac{\mathrm{d}M_y}{\mathrm{d}z}; \qquad T_\omega = -\frac{\mathrm{d}M_\omega}{\mathrm{d}z} \tag{1.34}$$

Substituting eqn (1.33) into (1.32) and using eqn (1.34) yields

$$\tau t = \frac{V_y Q_x}{I_{xx}} - \frac{V_x Q_y}{I_{yy}} + \frac{T_\omega Q_\omega}{\Gamma} \tag{1.35}$$

where

$$Q_x = \int_0^s yt\,\mathrm{d}s; \qquad Q_y = \int_0^s xt\,\mathrm{d}s; \qquad Q_\omega = \int_0^s (\bar{\omega}_s - \omega_s)t\,\mathrm{d}s \tag{1.36}$$

Note that when $s = m$ or $s = 0$, the three Q's are zero.

A strain associated with eqn (1.35) may now be formulated by dividing eqn (1.35) by Gt. The complementary strain energy is found by dividing eqn (1.35) by t, squaring, dividing by $2G$, and integrating over the volume. A typical square term, associated with V_y, is

$$\int_0^L \tfrac{1}{2}\frac{V_y^2}{G\lambda_y A}\,\mathrm{d}z \tag{1.36a}$$

where

$$\lambda_y = I_{xx}^2 A \bigg/ \int_0^m \frac{Q_x^2}{t}\,\mathrm{d}s \tag{1.36b}$$

Another use of complementary energy is associated with thin-walled *closed* sections. It is shown in elementary textbooks that the shear flow around a closed thin-walled section under a torque T is

$$\tau t = T/2A_s \tag{1.37}$$

where A_s is the area enclosed by the midline of the section. Again, it would be difficult to infer this directly from strain considerations. The complementary strain energy associated with eqn (1.37) is

$$\int_0^L \left\{\tfrac{1}{2}\frac{T^2}{(2A_s)^2 G}\oint_c \frac{1}{t}\,\mathrm{d}s\right\}\mathrm{d}z \tag{1.38}$$

With closed sections, warping is generally not considered.

Other variational principles may be derived as a combination of eqns (1.1) and (1.2) or by adding various constraints to the formulation with associated Lagrange multipliers. For instance, suppose it was desired that v and ϕ_x were to be thought of as 'independent', but were related by the constraint in eqn (1.4). Equation (1.4) should be rearranged into a homogeneous form, multiplied by a Lagrange multiplier and integrated from 0 to L and added to the potential energy functional. It would be seen that the multiplier was physically equivalent to V_y. In fact, if the rearranged eqn (1.4) is not homogeneous, then v and ϕ_x are independent and $dv/dz + \phi_x$ represents a good strain measure on which to build a potential energy term due to bending shear.

1.3. FINITE ELEMENTS AVAILABLE FOR NON-CURVED THIN-WALLED SECTIONS

The discussion thus far has been to describe the force and displacement types common to non-curved thin-walled structural members, and to express the formulation in terms of energy principles. The energy principles serve as a criterion for comparing, and even legitimizing, approximate solutions. The finite element method is a set of procedures that uses the energy methods to formulate approximate solutions. The set of procedures can be summarized as follows:

(1) Discretize the volume of the structure into subvolumes, known as finite elements.
(2) For a particular element, choose an appropriate energy principle. Choose an admissible field form with undetermined parameters appropriate for that principle. Not only must the continuity requirements be fulfilled across element boundaries, but a state of constant strain/stress must be possible in order that the strain/stress can approach a non-zero value as the element size shrinks to zero.
(3) Using the field in the energy principle, form element equations, generally, between nodal forces and displacements.
(4) Assemble the equations to form global equations relating the undetermined parameters with the external loading. Employ the structure's boundary conditions.
(5) After solving the global equations, trace back through the procedure to calculate stresses and strains for each element.

Step (1) of course, is problem dependent and requires a combination of experience and common sense. For our purpose we are interested in structures such that all or part of the structure can be modeled as a non-curved thin-walled section.

If we have a thin-walled section, the previous discussion above indicates some choices for an energy principle. Much of the discussion in this section will discuss the rest of step (2) and step (3). Steps (4) and (5) are the most general in finite element analysis and will be discussed below.

Suppose an entire structure was modeled using the minimum potential energy principle. The strain energy U_i for the ith element will be shown to be expressible as

$$U_i = \tfrac{1}{2}\{a_i\}^{\mathrm{T}}[K_i]\{a_i\} \tag{1.39}$$

where $\{a_i\}$ are a set of parameters describing the displacement field choice $\mathbf{u}_i$ over the ith element, and $[K_i]$ is a derived matrix known as the element stiffness. The total strain energy for the structure is

$$U = \sum_{i=1}^{N} U_i = \sum_{i=1}^{N} \tfrac{1}{2}\{a_i\}^{\mathrm{T}}[K_i]\{a_i\} = \tfrac{1}{2}\{a\}^{\mathrm{T}}[K_a]\{a\} \tag{1.40}$$

where $\{a\}$ is formulated by listing all the $\{a_i\}$ sequentially in one long column vector, and thus $[K_a]$ is formulated by listing the individual $[K_i]$ along the main 'diagonal' of an otherwise zero matrix. The $\{a_i\}$'s are related to the nodal displacements $\{D\}$ by the compatibility condition

$$\{a\} = [C]\{D\} \tag{1.41}$$

Substituting eqn (1.41) into eqn (1.40) yields

$$U = \tfrac{1}{2}\{D\}^{\mathrm{T}}[C]^{\mathrm{T}}[K_a][C]\{D\} = \tfrac{1}{2}\{D\}^{\mathrm{T}}[K]\{D\} \tag{1.42}$$

where $[K]$ is called the global stiffness matrix. Suppose the vector $\{P\}$ represents nodal external forces; then the potential energy functional is expressed as

$$\Pi_{\mathrm{p}} = U - \{D\}^{\mathrm{T}}\{P\} \tag{1.43}$$

Taking variations with respect to $\{D\}$ yields the global stiffness equations

$$[K]\{D\} = \{P\} \tag{1.44}$$

Equation (1.44) is inverted, and with $\{D\}$ in hand, all quantities of interest may be solved for.

Now suppose that a structure was modeled using the minimum complementary energy principle. The complimentary strain energy U_i for the ith element may be expressed as

$$U_i^{\mathrm{T}} = \tfrac{1}{2}\{r_i\}^{\mathrm{T}}[f_i]\{r_i\} \tag{1.45}$$

where the $\{r_i\}$ are unknown element nodal forces and $[f_i]$ is the element flexibility matrix. The total complementary strain energy is thus

$$U^* = \sum_{i=1}^{N} U_i = \sum_{i=1}^{N} \tfrac{1}{2}\{r_i\}^{\mathrm{T}}[f_i]\{r_i\} = \tfrac{1}{2}\{r\}^{\mathrm{T}}[F_r]\{r\} \tag{1.46}$$

where $\{r\}$ and $[F_r]$ are formed from the $\{r_i\}$'s and $[f_i]$'s in a manner comparable to the way $\{a\}$ and $[K_a]$ are formed from the $\{a_i\}$'s and $[K_i]$'s. The $\{r\}$'s are related to $\{P\}$ by the equilibrium condition

$$[S]\{r\} = \{P\} \tag{1.47}$$

In general, $[S]$ is not square and contains many more columns than rows, and thus cannot be inverted. What is done is that $\{r\}$ is partitioned into two sets $\{r_0\}$ and $\{r_1\}$ so that eqn (1.47) is written as

$$[S_0]\{r_0\} + [S_1]\{r_1\} = \{P\} \tag{1.48}$$

where $[S_0]$ is square and invertible. Solving eqn (1.48) for $\{r_0\}$ yields

$$\begin{aligned}\{r\} = \begin{Bmatrix} r_0 \\ r_1 \end{Bmatrix} &= \begin{bmatrix} -[S_0]^{-1}[S_1] \\ [I] \end{bmatrix}\{r_1\} + \begin{bmatrix} [S_0]^{-1} \\ [0] \end{bmatrix}\{P\} \\ &= [D_1]\{r_1\} + [D_2]\{P\}\end{aligned} \tag{1.49}$$

If the complementary energy functional is expressed as

$$\Pi_c = U^* - \{P\}^{\mathrm{T}}\{D\} \tag{1.50}$$

then, after substituting eqn (1.49) into eqn (1.46), eqn (1.50) becomes

$$\begin{aligned}\Pi_c = {} & \tfrac{1}{2}\{r_1\}^{\mathrm{T}}[\phi_{11}]\{r_1\} + \tfrac{1}{2}\{r_1\}^{\mathrm{T}}[\phi_{12}]\{P\} \\ & + \tfrac{1}{2}\{P\}^{\mathrm{T}}[\phi_{21}]\{r_1\} + \tfrac{1}{2}\{P\}^{\mathrm{T}}[\phi_{22}]\{P\} - \{P\}^{\mathrm{T}}\{D\}\end{aligned} \tag{1.51}$$

where

$$[\phi_{ij}] = [D_i]^{\mathrm{T}}[F_r][D_j], \qquad i, j = 1, 2 \tag{1.52}$$

Taking variations of eqn (1.51) with respect to $\{r_1\}$ and $\{P\}$ yields the two sets of equations

$$\begin{aligned}&[\phi_{11}]\{r_1\} + \tfrac{1}{2}([\phi_{12}] + [\phi_{21}]^{\mathrm{T}})\{P\} = \{0\} \\ &\tfrac{1}{2}([\phi_{12}]^{\mathrm{T}} + [\phi_{21}])\{r_1\} + [\phi_{22}]\{P\} = \{D\}\end{aligned} \tag{1.53}$$

The first of eqn (1.53) can be solved for $\{r_1\}$ and substituted into the second to yield the global flexibility relations:

$$[F]\{P\} = \{D\} \tag{1.54}$$

where

$$[F] = [\phi_{22}] - \tfrac{1}{4}([\phi_{12}]^{\mathrm{T}} + [\phi_{21}])[\phi_{11}]^{-1}([\phi_{12}] + [\phi_{21}]^{\mathrm{T}}) \tag{1.55}$$

These methods are rather general and are presented in an abstract form. They should become more clear as particular cases are worked out.

The above global procedures depend upon the derivation of the $[k_i]$'s or $[f_i]$'s. We now describe how these matrices are derived. At first, it will be fruitful to look at the various types of loading individually.

1.3.1. Axial Thrust

Figure 1.3 illustrates the axial thrust element. The z-axis is termed a 'local' axis, and is convenient to use when looking only at the element.

In section 1.2 it was pointed out that the boundary conditions for axial thrust consisted of one condition at each end of the bar, for a total of two independent conditions to be satisfied. This implies that at least two independent parameters be chosen to describe the displacement field w. Associated with these parameters are functions of z to be

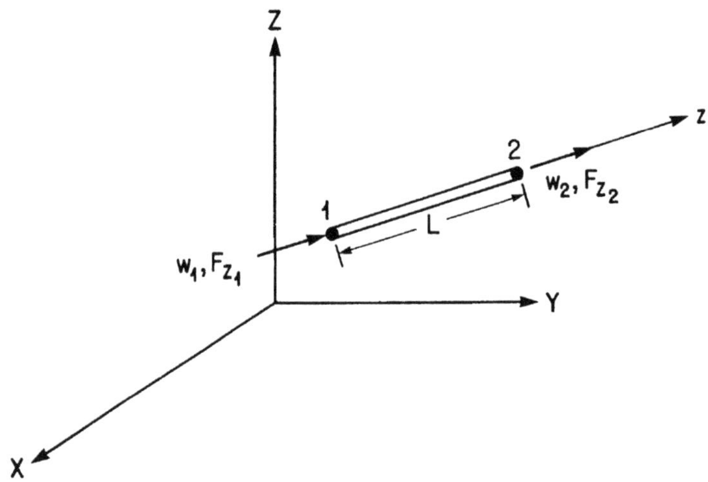

FIG. 1.3. Axial thrust element.

chosen. If the parameters are labeled a_1 and a_2, then let w be approximated as

$$w = a_1 f_1(z) + a_2 f_2(z) \tag{1.56}$$

We must choose f_1 and f_2 to satisfy some other conditions. First, we should include rigid body motion of the element. While this does not lead to stress or strain, it must be included because other parts of the structure will reposition this element whether this element deforms or not. Rigid body motion is represented by constant w. Thus, let $f_1(z)$, say, be the function 1. Secondly, we must allow for a constant state of strain. Since the strain measure is the first derivative of w, then we need a term proportional to z, so that when the derivative is taken a constant term appears. Thus, let $f_2(z) = z$. This second condition assures that w is continuous within the element up to the strain term. The third condition for admissibility is that w must be continuous from element to element. If we identify $w(0)$ with w_1 and $w(L)$ with w_2, and use an appropriate form of eqn (1.41), this will be accomplished with no extra terms. Thus, an admissible form of w is

$$w = a_1 + a_2 z \tag{1.57}$$

The a's may be determined in terms of w_1 and w_2 by the general relation

$$\{d\} = [A]\{a\} \tag{1.58}$$

where

$$\{d\} = \begin{Bmatrix} w_1 \\ w_2 \end{Bmatrix} \tag{1.59}$$

It is seen that

$$[A] = \begin{bmatrix} 1 & 0 \\ 1 & L \end{bmatrix} \tag{1.60}$$

Inverting eqn (1.58) and substituting into eqn (1.57) yields

$$w = (1 - z/L)w_1 + (z/L)w_2 = [1 - z/L \quad z/L]\begin{Bmatrix} w_1 \\ w_2 \end{Bmatrix} = [N]\{d\} \tag{1.61}$$

The matrix $[N]$ is referred to as the shape functions, and the collection $\{d\}$ as the element degrees-of-freedom.

The strain is the first derivative of w; thus,

$$\varepsilon = \frac{\mathrm{d}w}{\mathrm{d}z} = -\frac{1}{L}w_1 + \frac{1}{L}w_2 = [-1/L \quad 1/L]\begin{Bmatrix} w_1 \\ w_2 \end{Bmatrix} = [B]\{d\} \tag{1.62}$$

where $[B]$ is the strain–displacement matrix. For a linearly elastic rod, the stress is the strain times Young's modulus, or

$$\sigma = E\varepsilon = E[-1/L \quad 1/L]\begin{Bmatrix} w_1 \\ w_2 \end{Bmatrix} = [E][B]\{d\} = [k_\sigma]\{d\} \tag{1.63}$$

where $[k_\sigma]$ is referred to as the element stress matrix. The strain energy is now formulated as

$$\begin{aligned} U_1 &= \int_v \tfrac{1}{2}E\varepsilon^2 \,\mathrm{d}V = \int_0^L \tfrac{1}{2}\varepsilon^{\mathrm{T}}[E]\varepsilon A \,\mathrm{d}z \\ &= \tfrac{1}{2}\int_0^L \begin{Bmatrix} w_1 \\ w_2 \end{Bmatrix}^{\mathrm{T}} \begin{bmatrix} -1/L \\ 1/L \end{bmatrix} E[-1/L \quad 1/L]\begin{Bmatrix} w_1 \\ w_2 \end{Bmatrix} A \,\mathrm{d}z \\ &= \int_v \tfrac{1}{2}\{d\}^{\mathrm{T}}[B]^{\mathrm{T}}[E][B]\{d\}\,\mathrm{d}V \end{aligned} \tag{1.64}$$

Note that $\{d\}$ is just a set of pure numbers independent of the coordinate position, and thus may be taken outside the integration sign. Performing the integral in eqn (1.64) yields the element stiffness matrix $[k]$ as

$$[k] = \int_v [B]^{\mathrm{T}}[E][B]\,\mathrm{d}V = \frac{AE}{L}\begin{bmatrix} 1 & -1 \\ -1 & 1 \end{bmatrix} \tag{1.65}$$

If F_{z_1} and F_{z_2} are the resultants of the forces that other elements and external agents exert on the rod at points 1 and 2, respectively. Then the element stiffness equations are

$$[k]\{d\} = \{r\} \tag{1.66}$$

or

$$\frac{AE}{L}\begin{bmatrix} 1 & -1 \\ -1 & 1 \end{bmatrix}\begin{Bmatrix} w_1 \\ w_2 \end{Bmatrix} = \begin{Bmatrix} F_{z_1} \\ F_{z_2} \end{Bmatrix} \tag{1.67}$$

Note that $[k]$ is singular, with the order of singularity being equal to the number of rigid body motions available to the rod, in this case, 1. This implies that F_{z_1} and F_{z_2} would not be independent of each other for independent w_1 and w_2. In fact, their sum is zero, which is the appropriate equilibrium equation for this element.

If a force per unit length q_z acts on the rod, then its external virtual work would be

$$\int_0^L q_z \delta w \,\mathrm{d}z \tag{1.68}$$

Substituting eqn (1.61) into eqn (1.68) yields

$$\{\delta d\}^{\mathrm{T}}\{r_{\mathrm{ext}}\} = \begin{Bmatrix} \delta w_1 \\ \delta w_2 \end{Bmatrix}^{\mathrm{T}} \int_0^L [1 - z/L \quad z/L]^{\mathrm{T}} q_z \, \mathrm{d}z \tag{1.69}$$

Thus, $\{r_{\mathrm{ext}}\}$ may be added to the right-hand side of eqn (1.67). If q_z is not present, the linear shape functions of eqn (1.61) satisfy equilibrium in the interior of the element, while the global procedure can be shown to satisfy equilibrium at the nodal points (see the numerical examples). Thus, for rod elements under no distributed loads, such as in truss problems, we have hit upon the exact solution form. Further discretization of the structure or sophistication of the element is unnecessary. When q_z is present, however, the solution is not exact, and further discretization or level of sophistication will improve the solution.

If the rod is oriented at an angle relative to the global x-, y- and z-axes, then it is usually good practice to transform eqns (1.66) or (1.67) to global co-ordinates. If l, m and n are direction cosines of the z-axis related to the X-, Y- and Z-axes respectively, then

$$\begin{Bmatrix} w_1 \\ w_2 \end{Bmatrix} = \begin{bmatrix} l & m & n & 0 & 0 & 0 \\ 0 & 0 & 0 & l & m & n \end{bmatrix} \begin{Bmatrix} U_1 \\ V_1 \\ W_1 \\ U_2 \\ V_2 \\ W_2 \end{Bmatrix} \tag{1.70}$$

or, in general,

$$\{d_{\mathrm{local}}\} = [T]\{d_{\mathrm{global}}\} \tag{1.71}$$

Substituting in eqn (1.66) gives

$$\begin{aligned} [k_{\text{relative to global}}] &= [T]^{\mathrm{T}}[k_{\text{relative to local}}][T] \\ \{r_{\text{relative to global}}\} &= [T]^{\mathrm{T}}\{r_{\text{relative to local}}\} \end{aligned} \tag{1.72}$$

Having used the potential energy formulation, we can now discuss the complementary energy formulation. First, we note that rigid body motion does not result in stresses; thus, a good beginning choice for a particular element formulated using the complementary energy approach is to assume that the number of parameters describing the stress field is the number of degrees of freedom for the element minus the number of rigid body motions available for that element. In this case, this leaves one independent parameter. If a state of constant stress is to be possible, then

$$\sigma = \sigma_1, \quad \text{a constant} \tag{1.73}$$

is certainly a possible candidate. The choice of σ must satisfy the homogeneous equations of equilibrium. This constant state of stress certainly does. Finally, equilibrium must exist at the boundaries between elements. The nodal equilibrium equations are represented by eqn (1.47), and thus eqn (1.73) represents an admissible stress field. Introducing $P_1 = A\sigma_1$, in general,

$$\sigma = [Z]\{\sigma\} = \left[\frac{1}{A}\right]\{P_1\} \tag{1.74}$$

for this case. The strain is found as

$$\varepsilon = \frac{\sigma}{E} = \frac{P_1}{EA} = [E]^{-1}[Z]\{\sigma\} \tag{1.75}$$

Using eqns (1.74) and (1.75) in the complementary strain energy yields

$$U_1^* = \tfrac{1}{2}P_1^{\mathrm{T}}fP_1 = \tfrac{1}{2}\{\sigma\}^{\mathrm{T}}[f]\{\sigma\} \tag{1.76}$$

where

$$f = \frac{L}{EA} = \int_V [Z]^{\mathrm{T}}[E]^{-1}[Z]\,\mathrm{d}V \tag{1.77}$$

A relation exists between P_1 and F_{z_1} and F_{z_2}; this relation is

$$\begin{Bmatrix} F_{z_1} \\ F_{z_2} \end{Bmatrix} = \begin{bmatrix} -1 \\ 1 \end{bmatrix}\{P_1\} \tag{1.78}$$

or, in general,

$$\{r\} = [H]\{\sigma\} \tag{1.79}$$

This relation is substituted into eqn (1.47) and carried through the global assembly procedure with $\{\sigma\}$ as the basic unknowns.

If q_z is present, σ must satisfy the inhomogeneous equations. A particular solution σ_{p} is needed. If σ_{c} is the inhomogeneous stress field, then

$$\delta\sigma = \delta(\sigma_{\mathrm{c}} + \sigma_{\mathrm{p}}) = \delta\sigma_{\mathrm{c}} \tag{1.80}$$

However,

$$\varepsilon = \sigma/E = \frac{1}{E}(\sigma_c + \sigma_p) \tag{1.81}$$

The complementary strain energy thus contains a term

$$\int_v \frac{\sigma_p}{E}\,\delta\sigma_c\,\mathrm{d}V \tag{1.82}$$

which gives an equivalent displacement to the problem.

1.3.2. Bending

Figure 1.4 represents a straight beam element. The degrees of freedom are again based upon the requirements of the boundary conditions. Here, we see four independent parameters for uniaxial bending. Anticipating similar characteristics between beam and rod, we immediately guess

$$v = a_0 + a_1 z + a_2 z^2 + a_3 z^3 \tag{1.83}$$

and investigate its admissibility. First of all, the first two terms represent rigid body translation and rotation, respectively. The third term covers the possibility of constant strain. The choice of the number of degrees of freedom ensures proper continuity, this time of the first derivative as well as v itself; thus, v is admissible. The appropriate shape functions that would result in converting from $\{a\}$

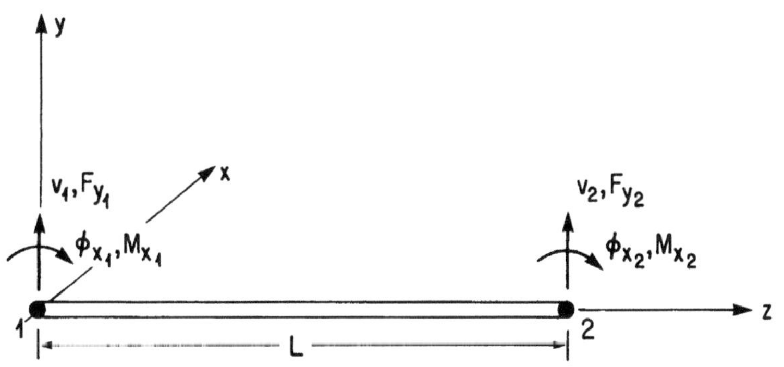

FIG. 1.4. Beam element.

to $\{d\}$ are

$$\begin{aligned} N_1 &= 1 - 3(z/L)^2 + 2(z/L)^3 \\ N_2 &= [-(z/L) + 2(z/L)^2 - (z/L)^3]L \\ N_3 &= 3(z/L)^2 - 2(z/L)^3 \\ N_4 &= [(z/L)^2 - (z/L)^3]L \end{aligned} \tag{1.84}$$

so that

$$v = [N_1 \;\; N_2 \;\; N_3 \;\; N_4]\begin{Bmatrix} v_1 \\ \phi_{x_1} \\ v_2 \\ \phi_{x_2} \end{Bmatrix} \tag{1.85}$$

The appropriate strain measure is the second derivative, with M_x the appropriate stress measure. The stiffness matrix is

$$[k] = \begin{bmatrix} 12EI_{xx}/L^3 & -6EI_{xx}/L^2 & -12EI_{xx}/L^3 & -6EI_{xx}/L^2 \\ -6EI_{xx}/L^2 & 4EI_{xx}/L & 6EI_{xx}/L^2 & 2EI_{xx}/L \\ -12EI_{xx}/L^3 & 6EI_{xx}/L^2 & 12EI_{xx}/L^3 & 6EI_{xx}/L^2 \\ -6EI_{xx}/L^2 & 2EI_{xx}/L & 6EI_{xx}/L^2 & 4EI_{xx}/L \end{bmatrix} \tag{1.86}$$

This matrix has a singularity of order 2. It also indicates that the nodal forces must satisfy the two equilibrium equations appropriate to a beam.

Like the axial thrust element, the beam element as formulated gives exact results when there is no distributed loading. When distributed loading is present the cubic polynomial is not exact and equilibrium will not be satisfied at every point.

The flexibility matrix can be formulated by first noting that a linear moment field is admissible. In particular, cutting the beam at an arbitrary position z and taking a free body diagram yields

$$M_x = -M_{x_1} - F_{y_1}z \tag{1.87}$$

Using (1.87) in the complementary energy function yields the flexibility equations

$$\begin{Bmatrix} V_1 \\ \phi_{x_1} \end{Bmatrix} = \begin{bmatrix} L^3/3EI_{xx} & L^2/2EI_w \\ L^2/2EI_{xx} & L/EI_{xx} \end{bmatrix}\begin{Bmatrix} F_{y_1} \\ M_{x_1} \end{Bmatrix} \tag{1.88}$$

If, say, $v_2 = \phi_{x_2} = 0$, appropriate stiffness equations may be derived by inverting eqn (1.88). Incorporating bending shear could be accomplished by adding a term (eqn (1.36a)) to the functional. Note that eqn

(1.87) implies $V_y = -F_y$, and thus the upper left-hand corner term is adjusted by adding $L/\lambda GA$. Inverting then would give the proper reduced stiffness.

1.3.3. Torsion

Figure 1.5 depicts an open thin-walled torsion element with appropriate degrees of freedom as specified by the boundary conditions. Note the introduction of $\phi_\omega = \mathrm{d}\phi_z/\mathrm{d}z$ and M_ω as an applied load. T_1 and T_2 are the total torques at nodes 1 and 2 respectively. Like the beam, it can successfully be argued that a cubic displacement field is admissible for ϕ_z. The element stiffness equations are

$$\begin{bmatrix} \frac{12E\Gamma}{L^3}+\frac{6}{5}\frac{GJ}{L} & \frac{6E\Gamma}{L^2}+\frac{1}{10}GJ & -\frac{12E\Gamma}{L^3}-\frac{6}{5}\frac{GJ}{L} & \frac{6E\Gamma}{L^2}+\frac{1}{10}GJ \\ \frac{6E\Gamma}{L^2}+\frac{1}{10}GJ & \frac{4E\Gamma}{L}+\frac{2}{15}GJL & -\frac{6E\Gamma}{L^3}-\frac{1}{10}GJ & \frac{2E\Gamma}{L}-\frac{1}{30}GJL \\ \frac{-12E\Gamma}{L^3}-\frac{6}{5} & \frac{GJ}{\mathrm{L}}-\frac{6E\Gamma}{L^2}-\frac{1}{10}GJ & \frac{12E\Gamma}{L^3}+\frac{6}{5}\frac{GJ}{L} & -\frac{6E\Gamma}{L^2}-\frac{1}{10}GJ \\ \frac{6E\Gamma}{L^2}+\frac{1}{10}GJ & \frac{2E\Gamma}{L}-\frac{1}{30}GJL & -\frac{6E\Gamma}{L^2}-\frac{1}{10}GJ & \frac{4E\Gamma}{L}+\frac{2}{15}GJL \end{bmatrix}$$

$$\times \begin{Bmatrix} \phi_{z_1} \\ \phi_{\omega_1} \\ \phi_{z_2} \\ \phi_{\omega_2} \end{Bmatrix} = \begin{Bmatrix} T_1 \\ M_{\omega_1} \\ T_2 \\ M_{\omega_2} \end{Bmatrix} \quad (1.89)$$

The stiffness matrix has an order of singularity of one. The only equilibrium equation implied by this matrix is that T_1 and T_2 add to zero.

Unlike the axial thrust or beam elements, the torsion element does not yield exact results even when there are no torques per unit length applied to the element. The cubic displacement field does not satisfy the equilibrium equations. The actual solution replaces the quadratic and cubic terms with hyperbolic functions. These functions, of course, satisfy all the criteria; however, we must differentiate between finite elements and what is referred to as matrix structural analysis. The first of these methods is meant to approximate, and therefore, is guessed

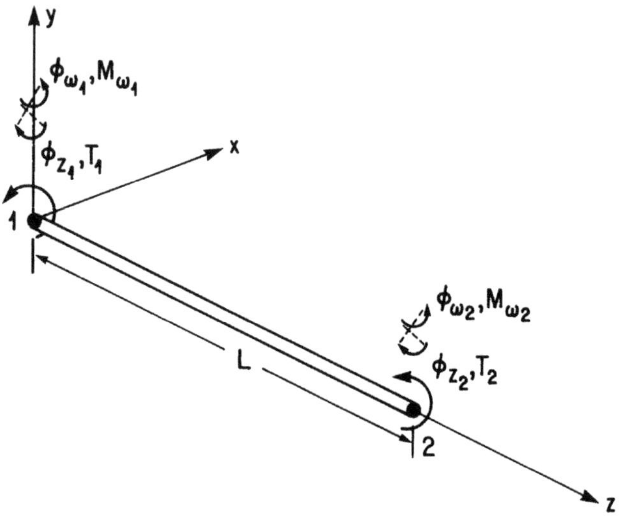

FIG. 1.5. Torsion element.

and tested for admissibility; the second is an exact method based upon the solution of the differential equations. If the hyperbolic functions were guessed correctly, the solution would be exact. Generally, most elementary finite element analyses use polynomials.

Cutting the member and taking a free body diagram indicates that the torque on a cross-section is

$$T = -T_1 \tag{1.90}$$

This, of course, is made up of two parts, the warping and St. Venant torques. The St. Venant torque may be expressed as

$$T_{sv} = T + \frac{dM_\omega}{dz} \tag{1.91}$$

If eqn (1.91) is used in eqn (1.30) or eqn (1.31), then M_ω must be considered continuous. We can then choose M_ω as

$$M_\omega = M_{\omega_1}(1 - z/L) + M_{\omega_2}(z/L) \tag{1.92}$$

Substituting into eqn (1.31) yields the flexibility equations

$$\begin{bmatrix} \dfrac{L}{GJ} & \dfrac{1}{GJ} & -\dfrac{1}{GJ} \\ \dfrac{1}{GJ} & \dfrac{1}{GJL}+\dfrac{L}{3E\Gamma} & -\dfrac{1}{GJL}+\dfrac{L}{6E\Gamma} \\ -\dfrac{1}{GJ} & -\dfrac{1}{GJL}+\dfrac{L}{6E\Gamma} & \dfrac{1}{GJL}+\dfrac{L}{3\Gamma} \end{bmatrix} \begin{Bmatrix} T_1 \\ M_{\omega_1} \\ M_{\omega_2} \end{Bmatrix} = \begin{Bmatrix} \phi_{z_1} \\ \phi_{\omega_1} \\ \phi_{\omega_2} \end{Bmatrix} \tag{1.93}$$

If warping shear effects are included (see eqn (1.35)), then the flexibility matrix is augmented by

$$\begin{bmatrix} 0 & 0 & 0 \\ 0 & 1/\lambda_\omega GI_pL & -1/\lambda_\omega GI_pL \\ 0 & -1/\lambda_\omega GI_pL & 1/\lambda_\omega GI_pL \end{bmatrix} \tag{1.94}$$

where I_p is the polar moment of area of the cross-section about the shear center and

$$\lambda = \left(\frac{\Gamma^2}{I_p}\right) \Big/ \int_0^m \frac{Q_\omega^2}{t}\,\mathrm{d}s \tag{1.95}$$

If the cross-section is closed, eqn (1.38) is used for the complementary energy. Equation (1.90) is enough to model the torque field, and the formulation of the flexibility matrix is comparable to that of the axial thrust element.

1.3.4. Combined Loading

Figure 1.6 represents the general non-curved thin-walled section. Note that there are seven degrees of freedom per node for a total of 14 degrees of freedom and their corresponding force quantities. If principal axes are chosen, then the 14×14 stiffness matrix can be broken into four square submatrices located about the main diagonal, and zeros everywhere else, because the individual loadings (thrust, bending about the x-axis, bending about the y-axis, torsion) are separable when the material is linear elastic. Schematically, the

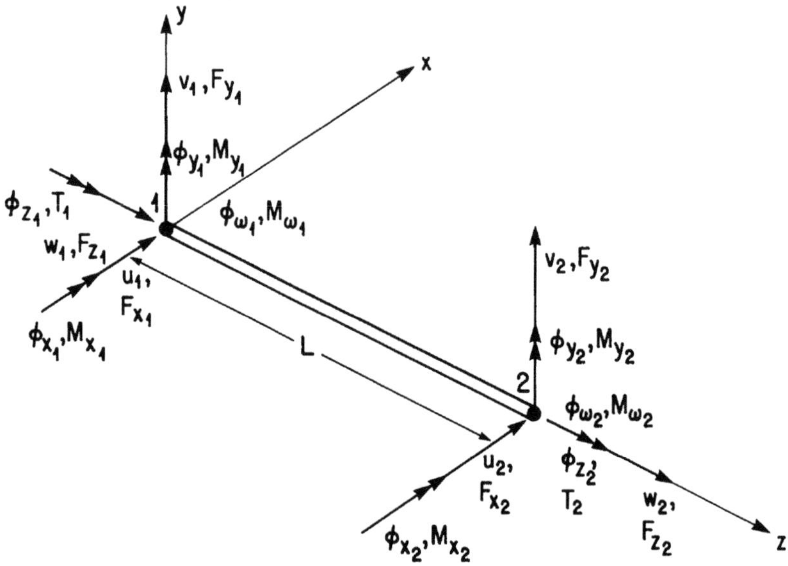

FIG. 1.6. General non-curved thin-walled section element.

element stiffness equations are as follows:

$$\left[\begin{array}{c|c|c|c} \text{eqn (1.65)} & 0 & 0 & 0 \\ \hline 0 & \text{eqn (1.86)} & 0 & 0 \\ \hline 0 & & \text{similar to eqn (1.86)} & 0 \\ \hline 0 & 0 & 0 & \text{eqn (1.89)} \end{array}\right] \begin{Bmatrix} w_1 \\ w_2 \\ v_1 \\ \phi_{x_1} \\ v_2 \\ \phi_{x_2} \\ u_1 \\ \phi_{y_1} \\ u_2 \\ \phi_{y_2} \\ \phi_{z_1} \\ \phi_{\omega_1} \\ \phi_{z_2} \\ \phi_{\omega_2} \end{Bmatrix} = \begin{Bmatrix} F_{z_1} \\ F_{z_2} \\ F_{y_1} \\ M_{x_1} \\ F_{y_2} \\ M_{x_2} \\ F_{x_1} \\ M_{y_1} \\ F_{x_2} \\ M_{y_2} \\ T_1 \\ M_{\omega_1} \\ T_2 \\ M_{\omega_2} \end{Bmatrix} \tag{1.96}$$

Equations (1.96) are the basic element equations for linear elastic straight thin-walled sections.

1.4. STABILITY CONSIDERATIONS

In this section we will consider the buckling of straight thin-walled sections under compression and/or pure uniaxial or biaxial bending. The phenomenon of buckling may be explained in various ways and formulated in many others, but the inclusion of non-linear terms in the strain–displacement law will lead to sets of equations that may not have unique solutions. Generally, a 'conventional' solution always seems to be present but a possible, different solution may be available when the applied load takes on specific values, or a specific range of values. The smallest such value that yields a non-trivial second solution is known as the critical (or buckling or bifurcation) load for the structure.

For our thin-walled member, instead of the linear relation (1.16a) for the normal strain, let us use

$$\varepsilon = \frac{\mathrm{d}W}{\mathrm{d}z} + \tfrac{1}{2}\left(\frac{\mathrm{d}U}{\mathrm{d}z}\right)^2 + \tfrac{1}{2}\left(\frac{\mathrm{d}V}{\mathrm{d}z}\right)^2 \tag{1.97}$$

where U, V and W are the displacements in the x, y and z directions of any point in the structural member. These displacements are related to u, v and w, the displacements of the centerline, and ϕ_z, the rotation about a z-axis running through the shear center, by

$$\begin{aligned} U &= u - (y - y_s)\phi_z; \qquad V = v + (x - x_s)\phi_z \\ W &= w - x\frac{\mathrm{d}u}{\mathrm{d}z} - y\frac{\mathrm{d}v}{\mathrm{d}z} + (\bar{\omega}_s - \omega_s)\frac{\mathrm{d}\phi_z}{\mathrm{d}z} \end{aligned} \tag{1.98}$$

where (x_s, y_s) are the coordinates of the shear center. Thus, it is seen that the W term leads to the normal strain used in the linear theory.

In general, a buckling problem may be formulated by assuming that the displacement $\mathbf{u}$ is partitioned as

$$\mathbf{u} = \mathbf{u}_0 + \xi\mathbf{u}_1 \tag{1.99}$$

where $\mathbf{u}_1$ is the normalized buckling mode, $\mathbf{u}_0$ is the fundamental solution and ξ is a parameter which can be assumed to be small. The buckling mode is linearly independent of the fundamental solution.

The non-linear strain displacement law

$$\boldsymbol{\varepsilon} = \mathbf{f}(\mathbf{u}) \tag{1.100}$$

may be expanded into a Taylor series about $\mathbf{u}_0$ as

$$\boldsymbol{\varepsilon} = \mathbf{f}(\mathbf{u}_0) + \xi\mathbf{f}'(\mathbf{u}_0)\mathbf{u}_1 + \tfrac{1}{2}\xi^2 f''(\mathbf{u}_0)\mathbf{u}_1^2 + \ldots \tag{1.101}$$

where $\mathbf{f}', \mathbf{f}'', \ldots$ are operators which are functions of $\mathbf{u}_0$ only. Equation (1.101) can be rewritten as

$$\boldsymbol{\varepsilon} = \boldsymbol{\varepsilon}_0 + \xi\boldsymbol{\varepsilon}_1 + \xi^2\boldsymbol{\varepsilon}_2 + \ldots \tag{1.102}$$

where it must be remembered that $\boldsymbol{\varepsilon}_j$ is a polynomial in $\mathbf{u}_1$ (or its partial derivatives) of order j. Applying the constitutive law, the stress field is seen to be in the form

$$\boldsymbol{\sigma} = \boldsymbol{\sigma}_0 + \xi\boldsymbol{\sigma}_1 + \xi^2\boldsymbol{\sigma}_2 + \ldots \tag{1.103}$$

where

$$\boldsymbol{\sigma}_j = [E]\boldsymbol{\varepsilon}_j, \qquad j = 1, 2, \ldots . \tag{1.104}$$

and thus $\boldsymbol{\sigma}_j$ is a polynomial in $\mathbf{u}_1$ of order j. Substituting eqns (1.99), (1.102) and (1.103) into the principle of virtual work (eqn (1.1)), and collecting terms of like powers of ξ yields

$$\begin{aligned}&\boldsymbol{\sigma}_0 * \delta\boldsymbol{\varepsilon}_0 + \xi(\boldsymbol{\sigma}_1 * \delta\boldsymbol{\varepsilon}_0 + \boldsymbol{\sigma}_0 * \delta\boldsymbol{\varepsilon}_1)\\ &\quad + \xi^2(\boldsymbol{\sigma}_2 * \delta\boldsymbol{\varepsilon}_0 + \boldsymbol{\sigma}_1 * \delta\boldsymbol{\varepsilon}_1 + \boldsymbol{\sigma}_0 * \delta\boldsymbol{\varepsilon}_2) + \ldots = \mathbf{F} * \delta\mathbf{u}_0 + \xi\mathbf{F} * \delta\mathbf{u}_1\end{aligned} \tag{1.105}$$

Equation (1.105) holds true for all ξ when like-powered terms on either side are equal. The zeroth order term defines the (non-linear) fundamental solution. Generally, the linear solution is taken to be sufficient. The first-order term generally is zero by the orthogonality of $\mathbf{u}_0$ and $\mathbf{u}_1$. Finally, if variations are taken with respect to $\mathbf{u}_1$, the second-order term gives the buckling mode equations, as

$$\boldsymbol{\sigma}_1 * \delta\boldsymbol{\varepsilon}_1 + \boldsymbol{\sigma}_0 * \delta\boldsymbol{\varepsilon}_2 = 0 \tag{1.106}$$

Note that the $\delta\boldsymbol{\varepsilon}_0$ term drops out because $\boldsymbol{\varepsilon}_0$ is independent of $\mathbf{u}_1$. Note further that eqn (1.106) is quadratic in $\mathbf{u}_1$, and thus the variational problem specified by eqn (1.106) will lead to linear, *homogeneous* equations. These equations will have non-trivial solutions only for certain values of the load parameter, of which $\boldsymbol{\sigma}_0$ is a function. The smallest non-zero value of the load parameter yielding non-trivial solutions is the buckling load.

Suppose the member is in compression with load P; then, the fundamental solution would yield

$$\sigma_0 = -P/A; \qquad w_0(z) = -Pz/EA \tag{1.107}$$

The buckling mode may be in the form of a u, v or ϕ_z displacement. From eqn (1.97) it is seen that

$$\left.\begin{aligned}
\varepsilon_1 &= \frac{\mathrm{d}W}{\mathrm{d}z} = -x\frac{\mathrm{d}^2u}{\mathrm{d}z^2} - y\frac{\mathrm{d}^2v}{\mathrm{d}z^2} + (\bar{\omega}_s - \omega_s)\frac{\mathrm{d}^2\phi_z}{\mathrm{d}z^2} \\
\varepsilon_2 &= \tfrac{1}{2}\left[\left(\frac{\mathrm{d}U}{\mathrm{d}z}\right)^2 + \left(\frac{\mathrm{d}V}{\mathrm{d}z}\right)^2\right] \\
&= \tfrac{1}{2}\left\{\left(\frac{\mathrm{d}u}{\mathrm{d}z}\right)^2 + \left(\frac{\mathrm{d}v}{\mathrm{d}z}\right)^2 + [(x-x_s)^2 + (y-y_s)^2]\left(\frac{\mathrm{d}\phi_z}{\mathrm{d}z}\right)^2 \right. \\
&\qquad \left. -\,2(y-y_s)\frac{\mathrm{d}u}{\mathrm{d}z}\frac{\mathrm{d}\phi_z}{\mathrm{d}z} + 2(x-x_s)\frac{\mathrm{d}v}{\mathrm{d}z}\frac{\mathrm{d}\phi_z}{\mathrm{d}z}\right\}
\end{aligned}\right\} \tag{1.108}$$

Substituting eqns (1.108), (1.104), and the St. Venant shear strain eqn (1.16b) into eqn (1.106), and performing the integrals over the cross-section, yields

$$\begin{aligned}
\delta\tfrac{1}{2}\int_0^L \Bigg\{ & EI_{xx}\left(\frac{\mathrm{d}^2v}{\mathrm{d}z^2}\right) + EI_{yy}\left(\frac{\mathrm{d}^2u}{\mathrm{d}z^2}\right)^2 + E\Gamma\left(\frac{\mathrm{d}^2\phi_z}{\mathrm{d}z^2}\right)^2 + GJ\left(\frac{\mathrm{d}\phi_z}{\mathrm{d}z}\right)^2 \\
& - P\Bigg[\left(\frac{\mathrm{d}u}{\mathrm{d}z}\right)^2 + \left(\frac{\mathrm{d}v}{\mathrm{d}z}\right)^2 + \frac{I_{\mathrm{P}}}{A}\left(\frac{\mathrm{d}\phi_z}{\mathrm{d}z}\right)^2 \\
& + 2y_s\frac{\mathrm{d}u}{\mathrm{d}z}\frac{\mathrm{d}\phi_z}{\mathrm{d}z} - 2x_s\frac{\mathrm{d}v}{\mathrm{d}z}\frac{\mathrm{d}\phi_z}{\mathrm{d}z}\Bigg]\Bigg\}\,\mathrm{d}z = 0
\end{aligned} \tag{1.109}$$

The highest order derivative appearing in the additional terms that are multiplied by P is the first; thus, if eqn (1.109) is used as a basis for finite element development, the additional terms pose no new requirements for satisfying admissibility for u, v or ϕ_z and thus the same cubic Hermitian functions used for conventional analysis may be used here. The element stiffness equations are in the form

$$([k] - P[k_g])\{d\} = \{r\} \tag{1.110}$$

where $[k]$ is the combined conventional element stiffness matrix depicted (minus the 2×2 submatrix for the axial thrust effect) in eqn (1.96) and $[k_g]$, referred to as the geometric stiffness, is developed from the additional terms of eqn (1.109). This matrix is square and symmetric. The nodal forces $\{r\}$ are the unknown internal forces between elements. After global assembly and incorporation of boundary conditions, the system of equations are

$$([K]-P[K_g])\{D\}=\{0\} \tag{1.111}$$

Non-trivial solutions to eqn (1.111) exist only when the multiplying matrix is singular. This occurs when its determinant is zero; thus,

$$|[K]-P[K_g]|=0 \tag{1.112}$$

represents an equation for values of P yielding non-trivial solutions to eqn (1.111). The smallest non-zero root of eqn (1.112) is the buckling load.

If the cross-section of the thin-walled section is doubly symmetric, then the shear center and centroid coincide, the two mixed terms in eqn (1.109) drop out, and the buckling problem decouples into three independent problems for u, v and ϕ_z. The smallest overall value of P obtained from solution of the three individual problems is the buckling load, and of course, indicates the mode of buckling.

Suppose the member is doubly symmetric and is loaded by equal and opposite bending moments M_{0x}; then

$$\sigma_0=M_{0x}y/I_{xx} \tag{1.113}$$

The fundamental solution is the displacement field v. The buckling mode will be a combination of u and ϕ_z. Substituting eqn (1.113) for eqn (1.107) in eqn (1.106) yields

$$\delta\tfrac{1}{2}\int_0^L\left\{EI_{yy}\left(\frac{\mathrm{d}^2u}{\mathrm{d}z^2}\right)^2+E\Gamma\left(\frac{\mathrm{d}^2\phi_z}{\mathrm{d}z^2}\right)^2+GJ\left(\frac{d\phi_z}{\mathrm{d}z}\right)^2-2M_{0x}\frac{\mathrm{d}u}{\mathrm{d}z}\frac{\mathrm{d}\phi_z}{\mathrm{d}z}\right\}\mathrm{d}z=0 \tag{1.114}$$

The form of the global equations is similar to that of eqn (1.111), with M_{0x} replacing P.

1.5. INELASTIC BEHAVIOR

In deriving the principle of minimum potential energy, it was assumed that the material had linear elastic behavior. This led to the quadratic form of that principle and, the linear form of the resulting finite element equations. This linear form has the desirable aspects of being conducive to matrix notation and a number of efficient means of solution. These properties make the finite element method an excellent candidate for computer application. Unfortunately, many structures experience non-linear behavioral characteristics due to yielding, creep, fatigue, thermal cycling etc. Not only is the stress–strain relationship non-linear, but it is largely history-dependent. The theoretical and experimental research into understanding constitutive laws is being performed by many individuals. Our purpose here is not to choose a particular model, but to choose a form to which many of the models conform, and apply it to finite element methods for thin-walled structures. This form is expressed mathematically as

$$\mathring{\boldsymbol{\sigma}} = [E_T]\mathring{\boldsymbol{\varepsilon}} - \mathring{\boldsymbol{\tau}} \tag{1.115}$$

To many readers, the dots over the field quantities may bring to mind a time derivative. Indeed, $\mathring{\boldsymbol{\sigma}}$ and $\mathring{\boldsymbol{\varepsilon}}$ are referred to as the stress rate and strain rate respectively; however, these rates may be taken with respect to a load parameter, rather than time, if the problem is otherwise independent of real time. The tangent modulus $[E_T]$ and the quantity $\mathring{\boldsymbol{\tau}}$ are in general, functions of stress, inelastic strain, temperature and material parameters, but are independent of the rate quantities; thus, eqn (1.115) represents linear equations in the rate quantities themselves.

The solution to inelastic analysis problems is similar to the solution of dynamics problems. From initial conditions we march in time, solving for increments of displacement and stress in each time interval and numerically integrating to obtain the time-history of the structure. The basis for solution of the rate problem is the principle of virtual work for the rate quantities, which is simply expressed as

$$\mathring{\boldsymbol{\sigma}} * \delta\mathring{\boldsymbol{\varepsilon}} = \mathring{\mathbf{F}} * \delta\mathring{\mathbf{u}} \tag{1.116}$$

Substituting eqn (1.115) into eqn (1.116) gives

$$[E_T]\mathring{\boldsymbol{\varepsilon}} * \delta\mathring{\boldsymbol{\varepsilon}} = \mathring{\mathbf{F}} * \delta\mathring{\mathbf{u}} + \mathring{\boldsymbol{\tau}} * \delta\mathring{\boldsymbol{\varepsilon}} \tag{1.117}$$

At any instant in time, $[E_T]$, $\mathring{\boldsymbol{\tau}}$ and $\mathring{\mathbf{F}}$ are known, thus leaving eqn (1.117) as a set of equations to be used for solving for $\mathring{\mathbf{u}}$.

Thin-walled structures present a special problem for this sort of formulation because, typically, the stress and strain measures commonly used (e.g. moment and curvature) are not ordinary stresses and strains as implied in eqn (1.115); and moment–curvature relationships, for example, are not readily apparent.

To retain the characteristics of the thin-walled section, we maintain eqn (1.16) as the strain–displacement laws. It must be remembered that except for the St. Venant shear strain, the other strain measures and their relation to the total strain are derived by geometric arguments, independent of the material properties. (As for the St. Venant shear, eqn (1.16b) is presumed to hold in inelastic cases as well. This approximation improves with the thinness of the section.) Thus, if the strain measures are derivatives of u, v, w and ϕ_z, the strain at any point in the cross-section can be expressed as

$$\boldsymbol{\varepsilon} = [Q]\mathbf{e} \tag{1.118}$$

where $\mathbf{e}$ represents the five strain measures (thrust, biaxial bending, warping, normal strain and St. Venant shear strain) and is related to the degrees of freedom $\{d\}$ by

$$\mathbf{e} = [B]\{d\} \tag{1.119}$$

and $[Q]$ is a function x, y (or n and s) appropriate for modeling eqn (1.16). Thus, the strain could be evaluated at any point in the thin-walled member by inserting the x and y co-ordinates into $[Q]$ and the z co-ordinate into $[B]$, if the $\{d\}$ were known.

We now discretize the thin-walled element itself into n longitudinal strips of cross-sectional area A_i located at point P_i in the cross-section (see ref. 2). The strain at the ith strip is a function of z, given as

$$\boldsymbol{\varepsilon}_i = [Q_i][B]\{d\} \tag{1.120}$$

Equation (1.120) may now be written in rate form and the constitutive law (eqn (1.115)) may be applied. For demonstration purposes here, let us assume that the $\mathring{\mathbf{t}} = 0$. The stress–strain law for the ith strip is written as

$$\mathring{\boldsymbol{\sigma}}_i = [E_T^i]\mathring{\boldsymbol{\varepsilon}}_i = [E_T^i][Q_i][B]\{\mathring{d}\} \tag{1.121}$$

where it must be remembered that $[E_T^i]$ is a function of stress, inelastic strain and temperature. In general, it will be a function of z and different for each strip i, and will more than likely be a full 2×2 rather than the diagonal form of the elastic case. The internal virtual work for a strip is

$$\delta U_i = \int_0^L \delta\mathring{\boldsymbol{\varepsilon}}_i^{\mathrm{T}}\boldsymbol{\sigma}_i A_i \,\mathrm{d}z = \{\delta\mathring{d}\}^{\mathrm{T}}\int_0^L [B]^{\mathrm{T}}[Q]^{\mathrm{T}}[E^i][Q_i][B]A_i \,\mathrm{d}z\{\mathring{d}\} \tag{1.122}$$

The integral in eqn (1.122) is identified as $[k_i]$, the strip stiffness, and is generally evaluated numerically. The total internal virtual work is found by summing eqn (1.122) over i; thus,

$$\delta U = \sum_{i=1}^{n} \delta U_i = \sum_{i=1}^{n} \{\delta \mathring{d}\}[k_i]\{\mathring{d}\} = \{\delta \mathring{d}\}\left(\sum_{i=1}^{n} [k_i]\right)\{\mathring{d}\} = \{\delta \mathring{d}\}[k]\{\mathring{d}\} \tag{1.123}$$

revealing the element stiffness. The external virtual work terms have not changed with the material behavior and thus are evaluated as before. This yields element stiffness equations in the familiar form. If the summation and integration are reversed, then the sum of eqn (1.122) is represented as

$$\delta U_i = \{\delta \mathring{d}\}^{\mathrm{T}} \int_0^L [B]^{\mathrm{T}} \left[\left(\sum_{i=1}^{n} [Q_i]^{\mathrm{T}} [E_T^i][Q_i] A_i \right) [B]\{\mathring{d}\} \right] \mathrm{d}z \tag{1.124}$$

The expression in the large brackets represents the relations between the strain measures $[B]\{\mathring{d}\}$ with the stress resultants.

The operation of summing over the strips is effective only when the strip areas A_i are chosen properly for a given set of cross-sectional locations P_i. A desirable property would be that the summation in the parentheses of eqn (1.124) reduce to the familiar EI_{xx}, EI_{yy}, EA, $E\Gamma$ and GJ for $[E_T] = [E_{\text{elastic}}]$. Figure 1.7 represents a typical discretization for a thin-walled section. Note that a double row of strips parallel to the section midline is indicated. This is to accommodate the St. Venant shear. The distance n away from the midline may be calculated

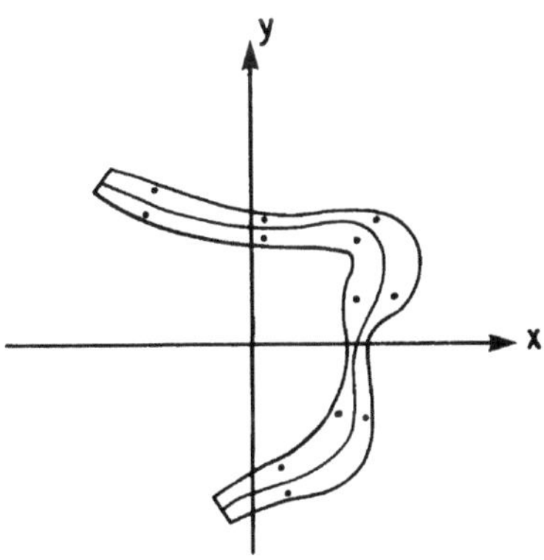

FIG. 1.7. Discretization of a cross-section of a thin-walled member to be used in non-linear analysis.

by relating a differential length's contribution to the St. Venant torque for these strips with the actual contribution for linearly elastic structures. With the shear strain given in eqn (1.16b), the shear stress is given as this strain times G; a differential force is calculated by multiplying by $\mathrm{d}A_i = (t_1/2)\,\mathrm{d}s$. On either side of the midline the force is equal and opposite, forming a couple with moment arm $2n^*$; thus, the differential moment is

$$\mathrm{d}M_{sv} = (2n^*)\left(2n^*G\frac{\mathrm{d}\phi_z}{\mathrm{d}z}\right)\left(\frac{t_i}{2}\mathrm{d}s\right) \tag{1.125}$$

We set this equal to

$$\mathrm{d}M_{sv} = \tfrac{1}{3}t_i^3\,\mathrm{d}s G\frac{\mathrm{d}\phi_z}{\mathrm{d}z} \tag{1.126}$$

thus yielding

$$n^* = t_i/\sqrt{6} = 0{\cdot}4082t_i \tag{1.127}$$

The s_i locations may be chosen arbitrarily. The A_i's, however, should satisfy

$$\begin{aligned}\sum_{i=1}^{n} A_i &= A\\ \sum_{i=1}^{n} y_i^2 A_i &= I_{xx}\\ \sum_{i=1}^{n} x_i^2 A_i &= I_{yy}\\ \sum_{i=1}^{n} (\bar{\omega}_s - \omega_{s_i})^2 A_i &= \Gamma\end{aligned} \tag{1.128}$$

where

$$\omega_{s_i} = \sum_{j=1}^{i} r_j s_j; \qquad \bar{\omega}_s = \sum_{i=1}^{n} \omega_{s_i}/n \tag{1.129}$$

In addition, we would want

$$\begin{aligned}\sum_{i=1}^{n} x_i A_i &= 0\\ \sum_{i=1}^{n} y_i A_i &= 0\end{aligned} \tag{1.130}$$

The six equations of (1.128) and (1.130) may be solved when six or more strips are used in the discretization. If more are used, additional equations such as

$$\sum_{i=1}^{n} x_i y_i A_i = 0 \qquad \text{(principal axes)}$$

$$\sum_{i=1}^{n} (\bar{\omega}_s - \omega_{s_i}) x_i A_i = 0 \tag{1.131}$$

$$\sum_{i=1}^{n} (\bar{\omega}_s - \omega_{s_i}) y_i A_i = 0$$

may be used.

While they do not contribute to the strain energy for members with large aspect ratio, the bending and warping shear stresses are present and may have to be calculated in order to approximately use the constitutive laws. This can be accomplished by using a numerical form for eqn (1.32); thus

$$\mathring{\boldsymbol{\tau}}_i = \frac{1}{t_i} \frac{\mathrm{d}}{\mathrm{d}z} \left(\sum_{j=1}^{i} \mathring{\boldsymbol{\sigma}}_j A_j \right) \tag{1.132}$$

1.6. NUMERICAL EXAMPLES

In this section numerical examples are given that demonstrate the points of previous sections.

Example 1: Thin-walled Section Under Torsion

Figure 1.8 depicts a thin-walled section, cantilevered at the end $z = 0$ and free at $z = L$, loaded by a torque T at that free end. One element will be used to model the structure. The purpose of this example is to demonstrate that the finite element method is only an approximate procedure. Once solved, the finite element solution will be compared to the exact solution for the more interesting quantities.

The global stiffness equations are found by first using the element stiffness equations (1.89) and recognizing that ϕ_{z_1} and ϕ_{ω_1} are zero by the boundary conditions, that $T_2 = T$ and that $M_{\omega_2} = 0$; thus,

$$\begin{bmatrix} \dfrac{12\,E\Gamma}{L^3} + \frac{6}{5}\dfrac{GJ}{L} & -\dfrac{6E\Gamma}{L^2} - \frac{1}{10}GJ \\ -\dfrac{6E\Gamma}{L^2} - \frac{1}{10}GJ & \dfrac{4E\Gamma}{L} + \frac{2}{15}GJL \end{bmatrix} \begin{Bmatrix} \phi_{z_2} \\ \phi_{\omega_2} \end{Bmatrix} = \begin{Bmatrix} T \\ 0 \end{Bmatrix} \tag{1.133}$$

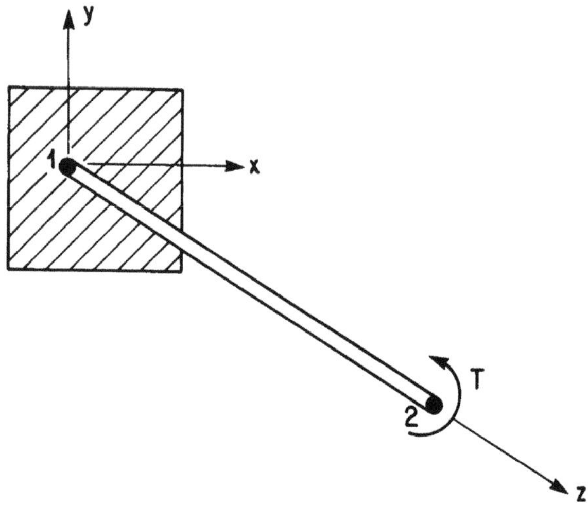

FIG. 1.8. Thin-walled section under torsion.

Solving for the degrees of freedom gives, in non-dimensional form,

$$\frac{GJ\phi_{z_2}}{TL}=\frac{4/\lambda^2+2/15}{12/\lambda^4+26/5\lambda^2+3/20}; \qquad \frac{GJ\phi_{\omega_2}}{T}=\frac{6/\lambda^2+1/10}{12/\lambda^4+26/5\lambda^2+3/20} \tag{1.134}$$

where

$$\lambda=\sqrt{\frac{GJ}{E\Gamma}}L \tag{1.135}$$

If the cross-sectional and material properties of the member are held fixed, then λ is indicative of the relative length of the member. Table 1.1 compares the solution (eqn (1.134)) and its consequences with the

TABLE 1.1

	$\lambda=0{\cdot}1$		$\lambda=1$		$\lambda=10$	
Quantity	*Exact*	*F.E.*	*Exact*	*F.E.*	*Exact*	*F.E.*
$GJ\phi_z(L)/TL$	0·00332	0·00332	0·2384	0·2382	0·9000	0·8530
$GJ\phi_z'(L)/T$	0·00498	0·00498	0·3519	0·3516	0·9999	0·7874
$E\Gamma''(0)/TL$	0·997	0·996	0·7616	0·7260	0·1000	0·0354
$-E\Gamma\phi'''(0)/T$	1	0·996	1	0·7488	1	0·0551

exact solution for the key structural quantities for three different values of λ, representing 'short', 'moderate' and 'long' members. For $\lambda \ll 1$, the approximation is excellent for all quantities. For $\lambda \approx 1$, the approximation worsens as further derivatives are taken, though the degrees of freedom are still approximated to a very high degree. For $\lambda \gg 1$, the results of the finite element analysis differ greatly with the exact solution. While this graphically demonstrates that the cubic Hermian element only approximates the exact solution, it implies that discretizing 'long' members into shorter ones with $\lambda < 1$ will provide adequate answers.

Example 2: Simple Truss Problem Solved by the Force Method

This example demonstrates the force-method of analysis. Figure 1.9 represents a simple, three-bar truss structure used in many optimization studies,[3] and in force-method research.[4] The loading and the supports are shown. A typical truss element is depicted in Fig. 1.10. As indicated in eqns (1.74)–(1.77), the bar element's stress field may be represented by a single parameter, its tension T, divided by its cross-sectional area. The flexibility of the element is its length divided by EA. The matrix relating T to the nodal forces is

$$[H] = \begin{Bmatrix} -\cos\theta \\ -\sin\theta \\ \cos\theta \\ \sin\theta \end{Bmatrix} \tag{1.136}$$

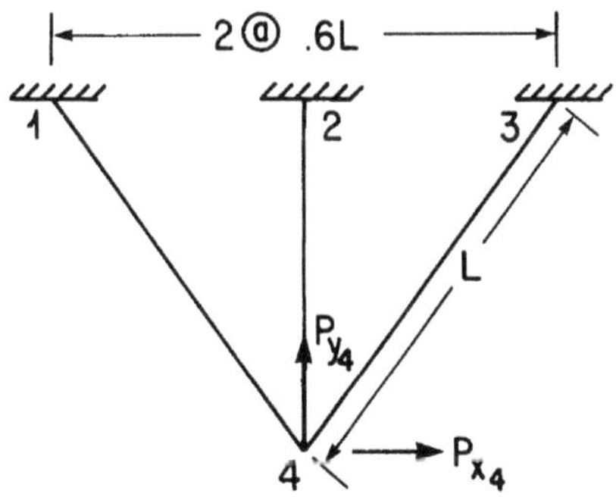

FIG. 1.9. Three-bar truss.

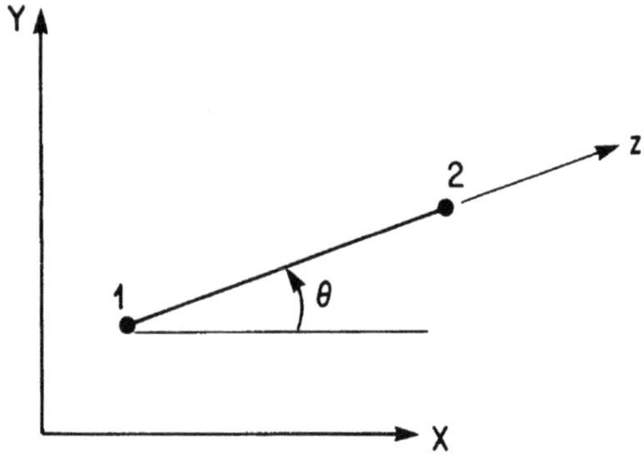

FIG. 1.10. Two-dimensional truss element.

The nodal equilibrium equations for this structure are easily determined as

$$\begin{Bmatrix} 0 \\ 0 \\ 0 \\ 0 \\ 0 \\ 0 \\ P_{x_4} \\ P_{y_4} \end{Bmatrix} = \begin{bmatrix} -3/5 & 0 & 0 & 1 & 0 & 0 & 0 & 0 & 0 \\ 4/5 & 0 & 0 & 0 & 1 & 0 & 0 & 0 & 0 \\ 0 & 0 & 0 & 0 & 0 & 1 & 0 & 0 & 0 \\ 0 & 1 & 0 & 0 & 0 & 0 & 1 & 0 & 0 \\ 0 & 0 & 3/5 & 0 & 0 & 0 & 0 & 1 & 0 \\ 0 & 0 & 4/5 & 0 & 0 & 0 & 0 & 0 & 1 \\ 3/5 & 0 & -3/5 & 0 & 0 & 0 & 0 & 0 & 0 \\ -4/5 & -1 & -4/5 & 0 & 0 & 0 & 0 & 0 & 0 \end{bmatrix} \begin{Bmatrix} T_1 \\ T_2 \\ T_3 \\ R_{x_1} \\ R_{y_1} \\ R_{x_2} \\ R_{y_2} \\ R_{x_3} \\ R_{y_3} \end{Bmatrix} \tag{1.137}$$

In order to systematically partition the matrix eqn (1.137) in the manner of eqn (1.48), a Gauss–Jordan elimination scheme with column pivoting is performed. If this $[S]$ matrix is augmented on the right with an 8×8 identity matrix, the row operations performed on $[S]$ lead to direct derivations of the $[D_1]$, and $[D_2]$ matrices. From a book-keeping point of view, it is usually advantageous to arrange $[D_1]$ and $[D_2]$ so that the original order of the elements of the $\{r\}$ vector is preserved. For this case the results of the procedure indicate that $\{r_1\}$, the set of redundants, is the single unknown T_2 and that:

$$[D_1] = \begin{bmatrix} -5/8 \\ 1 \\ -5/8 \\ -3/8 \\ 1/2 \\ 0 \\ -1 \\ 3/8 \\ 1/2 \end{bmatrix}; \quad [D_2] = \begin{bmatrix} 0 & 0 & 0 & 0 & 0 & 0 & 5/6 & -5/8 \\ 0 & 0 & 0 & 0 & 0 & 0 & 0 & 0 \\ 0 & 0 & 0 & 0 & 0 & 0 & -5/6 & -5/8 \\ 1 & 0 & 0 & 0 & 0 & 0 & 1/2 & -3/8 \\ 0 & 1 & 0 & 0 & 0 & 0 & -2/3 & 1/2 \\ 0 & 0 & 1 & 0 & 0 & 0 & 0 & 0 \\ 0 & 0 & 0 & 1 & 0 & 0 & 0 & 0 \\ 0 & 0 & 0 & 0 & 1 & 0 & 1/2 & 3/8 \\ 0 & 0 & 0 & 0 & 0 & 1 & 2/3 & 1/2 \end{bmatrix} \tag{1.138}$$

The submatrices within $[D_1]$ and $[D_2]$ outlined in the dashed brackets are those that will have an effect in the complementary strain energy computations. With

$$[F_r] = \frac{L}{EA} \begin{bmatrix} 1 & 0 & 0 \\ 0 & 4/5 & 0 \\ 0 & 0 & 1 \end{bmatrix} \tag{1.139}$$

the $[\phi_{ij}]$ matrices are

$$[\phi_{11}] = \frac{253}{160} \frac{L}{EA}; \qquad [\phi_{12}] = [\phi_{21}]^{\mathrm{T}} = [0 \;\; 25/32] \frac{L}{EA};$$

$$[\phi_{22}] = \begin{bmatrix} 25/18 & 0 \\ 0 & 25/32 \end{bmatrix} \frac{L}{EA} \tag{1.140}$$

Solving for T_2 yields

$$T_2 = \left(\frac{253}{160} \frac{L}{EA} \right)^{-1} [0 \;\; 25/32] \frac{L}{EA} \begin{Bmatrix} P_{x_4} \\ P_{y_4} \end{Bmatrix} = [0 \;\; -125/253] \begin{Bmatrix} P_{x_4} \\ P_{y_4} \end{Bmatrix} \tag{1.141}$$

The reaction forces and bar tensions may now be obtained. The results are

$$\begin{Bmatrix} T_1 \\ T_2 \\ T_3 \\ R_{x_1} \\ R_{y_1} \\ R_{x_2} \\ R_{y_2} \\ R_{x_3} \\ R_{y_3} \end{Bmatrix} = \begin{bmatrix} 5/6 & -80/253 \\ 0 & -125/253 \\ -5/6 & -80/253 \\ 1/2 & -48/253 \\ -2/3 & 64/253 \\ 0 & 0 \\ 0 & 125/253 \\ 1/2 & 48/253 \\ 2/3 & 64/253 \end{bmatrix} \begin{Bmatrix} P_{x_4} \\ P_{y_4} \end{Bmatrix} \tag{1.142}$$

To find the displacements, the global flexibility matrix is formed. The results are

$$\begin{Bmatrix} D_{x_4} \\ D_{y_4} \end{Bmatrix} = \frac{L}{EA} \begin{bmatrix} 25/18 & 0 \\ 0 & 100/253 \end{bmatrix} \begin{Bmatrix} P_{x_4} \\ P_{y_4} \end{Bmatrix} \tag{1.143}$$

This is the same answer as that obtained by using a 'classical' method of indeterminate analysis, as well as a finite element displacement method.

Example 3: Balcony Frame

This example makes use of the general thin-walled element as characterized by the stiffness matrix (eqn (1.96)). It also demonstrates how various effects are coupled through the enforcement of continuity of displacement after transforming from local to global co-ordinates.

Figure 1.11(a) shows a balcony frame consisting of three wide flange I-beams welded together to form right-angled corners. Figure 1.11(b) represents the finite element idealization of the half-frame, which due to symmetry is all that need be analysed.

If element 2 has its local z co-ordinate running from 2 to 3, then that element's local co-ordinate system is the same as the global system, and, except for a rearrangement of the order of the degrees of freedom, the stiffness of eqn (1.96) may be used directly for the element. If the local z-axis of element 1 runs from 1 to 2, then a

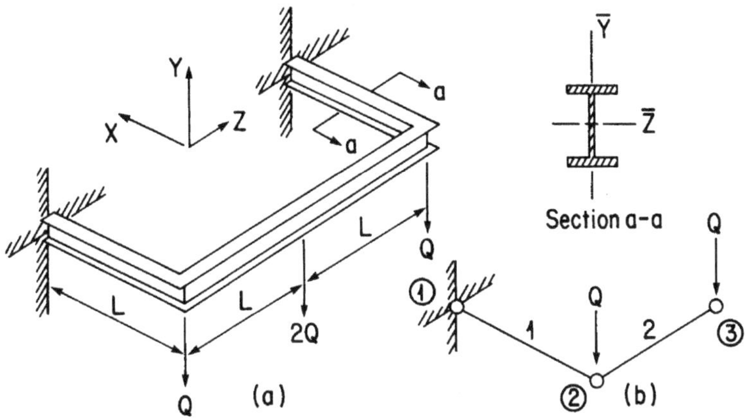

FIG. 1.11. (a) Balcony frame. (b) Finite element model for Fig. 1.11(a).

transformation of degrees of freedom becomes necessary. This transformation is:

$$\begin{Bmatrix} u \\ v \\ w \\ \phi_x \\ \phi_y \\ \phi_z \\ \phi_\omega \end{Bmatrix}_{\text{local}} = \begin{bmatrix} 0 & 0 & 1 & 0 & 0 & 0 & 0 \\ 0 & 1 & 0 & 0 & 0 & 0 & 0 \\ -1 & 0 & 0 & 0 & 0 & 0 & 0 \\ 0 & 0 & 0 & 0 & 0 & 1 & 0 \\ 0 & 0 & 0 & 0 & 1 & 0 & 0 \\ 0 & 0 & 0 & -1 & 0 & 0 & 0 \\ 0 & 0 & 0 & 0 & 0 & 0 & 0 \end{bmatrix} \begin{Bmatrix} u \\ v \\ w \\ \phi_x \\ \phi_y \\ \phi_z \\ \phi_\omega \end{Bmatrix}_{\text{global}} \tag{1.144}$$

It should be noted that ϕ_ω for these I-beams represents equal and opposite flange rotations about the y-axis, and thus the continuity of ϕ_ω is maintained at node 2. In general, if the connection was different or a more general cross-sectional shape was used, the continuity of ϕ_ω is not a straightforward condition to represent.

If the matrix in eqn (1.144) is derived as $[T]$, then its effect on the element stiffness equations is determined in the following manner. First, substitute eqn (1.144) into the potential energy functional

$$\Pi_p = \tfrac{1}{2}\{d\}^{\mathrm{T}}_{\text{local}}[k]_{\text{local}}\{d\}_{\text{local}} - \{d\}^{\mathrm{T}}_{\text{local}}\{r\}_{\text{local}} \tag{1.145}$$

to obtain

$$\pi_{\mathrm{p}} = \tfrac{1}{2}\{d\}^{\mathrm{T}}_{\mathrm{global}}([T]^{\mathrm{T}}[k]_{\mathrm{local}}[T])\{d\}_{\mathrm{global}} - \{d\}^{\mathrm{T}}_{\mathrm{local}}([T]^{\mathrm{T}})\{r\}_{\mathrm{local}} \tag{1.146}$$

From eqn (1.146) it is seen that

$$\begin{aligned} [k]_{\mathrm{global}} &= [T]^{\mathrm{T}}[k]_{\mathrm{local}}[T] \\ \{r\}_{\mathrm{global}} &= [T]^{\mathrm{T}}\{r\}_{\mathrm{local}} \end{aligned} \tag{1.147}$$

Thus, substituting eqn (1.96) and $[T]$ of eqn (1.144) into eqn (1.147) yields the element stiffness equations for element 1 with reference to the global degrees of freedom at nodes 1 and 2. This of course facilitates global assembly.

The boundary conditions are that all degrees of freedom at node 1 are zero, due to the cantilever end, and that w_3, ϕ_{x_3}, ϕ_{y_3} and ϕ_{ω_3} are zero due to the symmetry. The global stiffness equations are thus:

$$\begin{bmatrix} B_1+D & 0 & 0 & 0 & B_2 & 0 & 0 & -B_1 & 0 & 0 \\ & 2C_1 & 0 & -C_2 & 0 & C_2 & 0 & 0 & -C_1 & 0 \\ & & B_1+D & 0 & -B_2 & 0 & 0 & 0 & 0 & 0 \\ & & & C_3+A_1 & 0 & 0 & A_2 & 0 & C_2 & 0 \\ & & & & 2B_3 & 0 & 0 & -B_2 & 0 & 0 \\ & & & & & C_3+A_1 & A_2 & 0 & 0 & -A_1 \\ & & & & & & 2A_3 & 0 & 0 & -A_2 \\ & & & & & & & B_1 & 0 & 0 \\ \text{symmetrical} & & & & & & & & C_1 & 0 \\ & & & & & & & & & A_1 \end{bmatrix} \times \begin{Bmatrix} u_2 \\ v_2 \\ w_2 \\ \phi_{x_2} \\ \phi_{y_2} \\ \phi_{z_2} \\ \phi_{\omega_2} \\ u_3 \\ v_3 \\ \theta_{z_3} \end{Bmatrix} = \begin{Bmatrix} 0 \\ -Q \\ 0 \\ 0 \\ 0 \\ 0 \\ 0 \\ 0 \\ -Q \\ 0 \end{Bmatrix} \tag{1.148}$$

where

$$A_1 = \frac{12E\Gamma}{L^3} + \tfrac{6}{5}\frac{GJ}{L}; \qquad A_2 = \frac{6E\Gamma}{L^2} + \frac{GJ}{10};$$
$$A_3 = \frac{4E\Gamma}{L} + \tfrac{2}{15}GJL; \qquad D = \frac{AE}{L}$$
$$B_1 = \frac{12EI_{yy}}{L^3}; \qquad B_2 = \frac{6EI_{yy}}{L^2}; \qquad B_3 = \frac{4EI_{yy}}{L};$$
$$C_1 = \frac{12EI_{xx}}{L^3}; \qquad C_2 = \frac{6EI_{xx}}{L^2}; \qquad C_3 = \frac{4EI_{xx}}{L} \tag{1.149}$$

Using the following numerical data

$$A = 14{\cdot}4\ \text{in}^2; \quad I_{xx} = 272{\cdot}9\ \text{in}^4; \quad I_{yy} = 93{\cdot}0\ \text{in}^4;$$
$$J = 1{\cdot}38\ \text{in}^4; \quad \Gamma = 2070\ \text{in}^6; \quad L = 120\ \text{in}; \quad Q = 1000\ \text{lb} \tag{1.150}$$

the solution to eqn (1.148) is

$$\begin{Bmatrix} u_2 \\ v_2 \\ w_2 \\ \phi_{x_2} \\ \phi_{y_2} \\ \phi_{z_2} \\ \phi_{\omega_2} \\ u_3 \\ v_3 \\ \theta_{z_3} \end{Bmatrix} = \begin{Bmatrix} 0 \\ -0{\cdot}140\,712\ \text{in} \\ 0 \\ 0{\cdot}874\,746 \times 10^{-3}\ \text{rad} \\ 0 \\ 0{\cdot}175\,890 \times 10^{-2}\ \text{rad} \\ -0{\cdot}712\,574 \times 10^{-5}\ \text{rad in}^{-1} \\ 0 \\ -0{\cdot}210\,785\ \text{in} \\ 0{\cdot}142\,743 \times 10^{-2}\ \text{rad} \end{Bmatrix} \tag{1.151}$$

Example 4: Column Buckling

This example demonstrates the techniques developed in section 1.4 for buckling problems. Figure 1.12 illustrates the Euler column problem. We analyse half the column only due to the symmetry. The active degrees of freedom are ϕ_{x_1} and v_2. Buckling in the x direction or torsionally is assumed to be suppressed. Equations (1.111) take on the

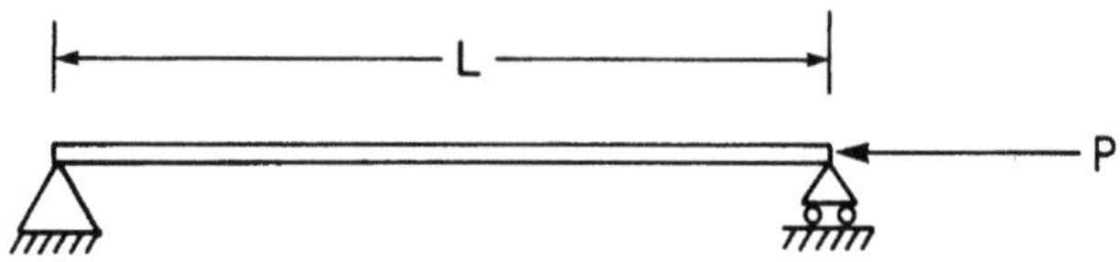

FIG. 1.12. Euler column problem.

form

$$\begin{bmatrix} 8\dfrac{EI_{xx}}{L} - \frac{1}{15}PL & -24\dfrac{EI_{xx}}{L^2} + \frac{1}{10}P \\ -24\dfrac{EI_{xx}}{L^2} + \frac{1}{10}P & 96\dfrac{EI_{xx}}{L^3} - \frac{12}{5}\dfrac{P}{L} \end{bmatrix} \begin{Bmatrix} \phi_{x_1} \\ v_2 \end{Bmatrix} = \begin{Bmatrix} 0 \\ 0 \end{Bmatrix} \quad (1.152)$$

Setting the determinant of the stiffness in eqn (1.152) to zero yields

$$\left(8\frac{EI_{xx}}{L} - \tfrac{1}{15}PL\right)\left(96\frac{EI_{xx}}{L^3} - \tfrac{12}{5}\frac{P}{L}\right) - \left(-24\frac{EI_{xx}}{L^2} + \tfrac{1}{10}P\right)^2 = 0 \quad (1.153)$$

The smallest root to eqn (1.153) is $PL^2/EI_{xx} = 9{\cdot}943\,85$. This is only 0·752% greater than the exact answer of π^2.

1.7. SUMMARY AND CONCLUSIONS

This chapter has laid the foundation for the development and use of finite elements for thin-walled structures. It has approached the development from both a displacement (stiffness) and force (flexibility) point of view, and has dealt principally with open sections because they exhibit the interesting warping effects. The chapter has developed methods for both stability and inelastic problem solution. A few numerical examples have been given to demonstrate some of the methods and principles discussed.

REFERENCES

1. SHAW, R. P., Unified finite element and boundary integral element methods, *6th Invitational Symposium on the Unification of Finite Elements, Finite Differences and Calculus of Variations*, 7 May 1982.

2. GELLIN, S., LEE, G. C. and CHERN, J. H., A finite element model for thin-walled members, *Int. J. Num. Meth. Engng,* **19,** 1983, 59–71.
3. BATT, J. R., GELLIN, S. and GELLATLY, R. A., *Force Method Optimization II, Vol. I—Theoretical Development,* AFWAL-TR-82-3088, November 1982.
4. GELLIN, S., Force method dynamics, *6th Invitational Symposium on the Unification of Finite Elements, Finite Differences and Calculus of Variations,* 7 May 1982.

Chapter 2

The Availability of Finite Element Software for Use with Thin-walled Structures

C. H. WOODFORD
Computing Laboratory, University of Newcastle upon Tyne, UK

2.1. INTRODUCTION

The finite element method is conceptually straightforward in that complex structures are idealised, in a representation, by a network or mesh of simpler interlocking structures, the simpler structures or finite elements being amenable to mathematical analysis. The analysis of the whole structure is obtained by simultaneously analysing the individual finite elements, having due regard to their individual positions within the mesh, and being totally dependent upon the assistance of an automatic computer. Even the most elementary of problems using a mesh made up of a few elements might require thousands of arithmetic operations and a sophisticated housekeeping strategy for the storing and retrieval of intermediate calculations. Even though such a problem will be solved within a matter of seconds on a modern computer, there are, at the other extreme, problems which require vast amounts of computer time. For example, a report[1] on the finite element analysis of a cylindrical body impacting normally against a stiffened cylindrical target, details a computer solution involving a total of 41 hours central processor time on a Cray Supercomputer. The success and general acceptance of the finite element method is due not only to its mathematical elegance but also to the way in which the method may be implemented on a modern computer. It can be shown[2] that to dissect for the purposes of analysis and then to reassemble for production of the results is, within limits, mathematically sound. The implementation is readily if somewhat painstakingly achieved by the use of well established programming languages such as FORTRAN, in any one of many operating environments. Full use may be made of generally available, reliable and robust hardware such as magnetic disks and tapes for the storage and retrieval of intermediates results. Modern interactive graphical display units lend themselves fully to the verification of data and the interpretation of results. The automation of the finite element method is an ideal application for modern computers, and can use every available computer facility fully.

2.2. THE STAGES OF FINITE ELEMENT ANALYSIS

Complete computer systems for finite element analysis have come to be constructed as a number of phases, executed in a manner that progresses from pre-processor phases, through the processor or solution phases and on to the final post-processor phases. A flow diagram of such a finite element scheme is shown in Fig. 2.1. The scheme shown does not represent a particular working system but it is sufficiently general to be used in the analysis of the various phases to be found in many of the commercially available systems for finite element analysis.

The pre-processor phase, Phase 1, consists of the input of data to specify the particular problem. Data may be supplied to the computer via a number of routes; for example, a keyboard, a bit pad, a mouse or possibly any of these in combination. A validation phase, Phase 2, in which the program attempts to identify any missing or conflicting data, follows. In the event of the detection of errors in Phase 2 a return is made to Phase 1 for the appropriate corrections to be made. In Phase 3, the system generates the finite element mesh based on the data supplied in Phase 1. Validation of the mesh takes place in Phase 4. If generation of the mesh involves deforming elements to an unacceptable extent an error is reported. The levels of acceptability or otherwise of element distortion will have been determined by either mathematical analysis and/or user experience. In cases of severe distortion it is advisable to return to Phase 1 so that the mesh may be redesigned. Indeed, the software may insist on this. The alternative is to accept the warnings made by the computer system and to treat subsequent results with extra caution. It has been found through long

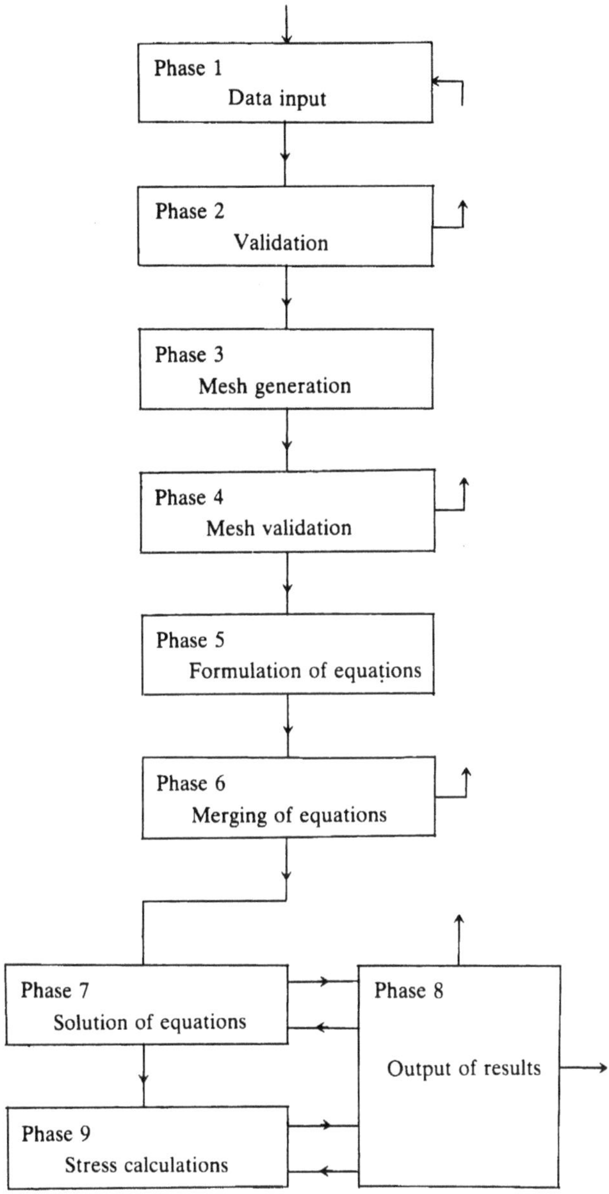

FIG. 2.1. Typical complete computer system for finite element analysis.

experience that the design of the mesh may critically affect the accuracy of the analysis. At Phase 4 the mesh might be displayed on a graphics device for approval. Ideally, there would be an interaction with the program to view the structure from a number of perspectives. Phase 4 concludes the pre-processor stages. The successful completion of Phase 4 implies that the computer can proceed to a possible solution of the problem. The finite element model, produced by the first four phases, may or may not be an accurate representation of the original problem, but that is decided by the program user and is quite beyond the scope of any present-day computer to decide. In addition, there is no guarantee that the problem as represented is solvable. It may transpire that some undetected discrepancy in the data renders the problem physically unrealistic, ambiguous, or even that as represented the problem is so inherently unstable as to defy solution using the available numerical techniques. Inherent computer 'round-off' error can make even well defined problems insolvable.

Phase 5 consists of the formulation of the linear equations, defining the individual properties of the individual elements in the finite element mesh and their relationship to their near neighbours. In Phase 6 the set of equations for each individual element is merged into one large system of equations. At this stage it may become apparent that the number of equations is not equal to the number of primary unknowns. This would preclude a unique solution and indicate that the original problem had not been specified either completely or unambiguously. These hitherto undetected data errors would necessitate a return to Phase 1. Conflicting specifications might produce an overdetermined system (more equations than unknowns) while an underspecification in the structure would produce an undetermined system (more unknowns than equations). Assuming that all is well, the resulting large system of equations will be of a banded form; the bandwidth being dependent upon the order in which the equations have been merged. The working storage required by the computer program for the solution of the large system of equations also depends upon the bandwidth, and varies with the bandwidth to the power of 2·5. In order to reduce the bandwidth to as near as possible to the theoretical minimum, the program may have a strategy for deciding, possibly in co-operation with the user, on which order to select the finite elements for generation of the linear equations. Apart from trying every possible combination, which might be prohibitively expensive in terms of computer resources, there appears to be no completely satisfactory algorithm for finding an optimum ordering within a reasonable efficiency. Even with an optimum bandwidth it might become apparent that the particular computing system still does not have sufficient working store available. In such a case the program would return to Phase 1, prompting the user to specify a less well refined mesh.

Having established the large system of equations, Phase 7 determines the primary unknowns which may typically be either displacements or temperatures. Having found the primary unknowns, the secondary unknowns, for example the stresses, are calculated in a later solution phase, Phase 9. The results available from Phase 7 and Phase 9 are voluminous and it is for this reason a separate post-processing stage is usually provided.

The post-processor stage, Phase 8, provides the user with the results actually required. Output may be, for example, in the form of printed tables, a generated report or graphical display. It is likely that the user will wish to interact with Phase 8 in order to consider all aspects of the results of the problem. In the light of Phase 8 the analysis may be accepted and a final fully documented report produced; perhaps a return to Phase 1 will be made for modifications, or the whole analysis may be laid aside for further consideration of the original problem and its abstraction as a finite element model. But, as with any computer modelling, there is the ever-present danger that attention is concentrated on the abstraction at the expense of the real problem. Time can so easily be wasted on tinkering with the computer model when the real problem lies in its original construction.

2.3. ESTABLISHING A COMPUTER PROGRAM: THEORY TO SOFTWARE APPROACH

There is a current divergence of views on how best to proceed to establishing a computer program that will solve a particular problem or range of problems. For clarity we will refer to the 'theory to software' approach and contrast it with the alternative approach of 'software to theory'. First, we discuss 'theory to software'. It might be considered desirable to construct a computer program following a pattern similar to that laid out in Fig. 2.1. At each stage some understanding of the theory would be necessary before proceeding to a computer implementation. Although this approach is intellectually appealing it does presuppose that a great deal of time is available for computer programming, testing and correction. Even if use is made of standard library procedures (for example the NAG routines[3] for the solution of a system of linear equations) the time and effort required from just one programmer might be prohibitive. In order to adopt this 'theory to software' approach it would be more realistic to incorporate the work of others in the field. A co-operative approach to what is in effect a considerable problem is a possible compromise solution, by drawing on existing work, yet at the same time allowing scope for individual self-expression and preserving the intellectual appeal.

There are a number of long-standing schemes which provide the basic software building blocks. By taking advantage of such a scheme the programmer is able to incorporate tried and tested routines within a self-constructed program. Not only is the user provided with reliable routines, but there is usually the encouragement to report on errors encountered, possible improvements or to contribute completely new routines. Such schemes require a constant supervisory and administrative effort for purposes of software documentation, correction, maintenance and development. The time and effort involved is considerable, and should not be underestimated by the programmer considering working in isolation on a scheme which it is hoped to make available to others. The programmer intent on developing a new system would be well advised to consider using the software building blocks as provided by any of the systems outlined in Appendix 1.

In the context of developing a program for the finite element method it is appropriate to sound a note of caution. Nothing is more likely to lead to disaster than trying to modify or use an existing program which has been handed down through successive generations of research workers. Software which is not properly maintained may rapidly deteriorate. Such programs, invariably written in some low-grade FORTRAN are recognisable by their DO LOOPS which often run remorselessly over several pages of source text. Comments are usually no more enlightening than 'C – N BECOMES N + 1' followed by 'N = N + 1' on the next line. Whatever N is and why it should be incremented is anyone's guess. The use of unwieldly COMMON blocks and of subroutines with inordinately long parameter lists are but two of the other less endearing features. Close inspection is likely to reveal a highly idiosyncratic and non-standard use of the language, based on a particular operating environment long since abandoned by the manufacturer. Not only is the author conspicuous by his absence, having long since departed for greener pastures, but so are the current users. To follow the logic behind a well annotated program is difficult enough. To understand any more than half a page of someone else's undocumented program (and for that matter one's own earlier undocumented programs) is well nigh impossible. To accept an unsupported program as a basis for future work is really to accept a poisoned cup. Far better and far less time-consuming is to begin with a blank sheet of paper.

2.4. ESTABLISHING A COMPUTER PROGRAM: SOFTWARE TO THEORY APPROACH

The alternative to the theory to software approach is to use one of the many general-purpose programs commercially available on a purchase, rental or bureau basis, the 'software to theory' approach. Large-scale programs represent the culmination of many man–years of work and provide a comprehensive approach to the solution of a wide range of problems. The highly competitive nature of the finite element industry ensures that products are under a constant state of review and generally take into account both user experience and the advantages of the latest computer technology. Particular attention has been paid to the pre- and post-processor stages to make the systems as user-friendly as possible. It is now quite common to be able to specify precisely what range of results are required and in what form. The process of presentation of data to the program and the subsequent automatic generation of the finite element mesh has been developed to such an extent that even the most complicated structure may be generated by just a few commands. There is every reason to suppose that this trend will be continued into providing an expert system environment enabling the user to converse directly with the program in his own engineering language, without having to translate into artificial pseudo-command languages, however simple they might be. There has been at least one previous, albeit unsuccessful, attempt. The SACON system[4] was constructed as a guide to users of the powerful MARC system (Appendix 2) for finite element analysis. By using a knowledge base concerning stresses and deflections of a structure under different loading conditions, SACON attempted to identify the required type of analysis and to recommend the most appropriate features of MARC to invoke. Its lack of success was due to the large number of necessary levels in the underlying SACON decision tree. However, it is not unreasonable to expect some future system, say SUPERSTRUT, to greet potential users with the message (no doubt suitably orchestrated): 'Hello, this is SUPERSTRUT. What's your problem?'.

Finite element systems implemented on work-stations and other such personal computers with graphical and printing facilities may be regarded as a design tool. Using such a tool it is possible to successively model a structure, perform the finite element analysis and then re-model until either a satisfactory solution is reached or the entire project is abandoned as being unrealistic. This whole process may be interactively supervised by the design engineer over a period ranging from hours to days and even weeks.

2.5. THE SOFTWARE TO THEORY OPTION

Given the all-embracing nature of the commercially available systems for finite element analysis it might appear eminently reasonable, having obtained assurances that the software performs the required analysis, to proceed directly to a solution and so benefit from the accumulation of many man–years of effort and then gratefully accept whatever results are produced. However, as with all tools and, particularly, complex design tools, it is essential to spend a considerable time in familiarisation with the mode of operation. It is advisable to tackle first a series of simple, textbook problems for which authenticated results are available. Only by this repeated series of tests will a degree of expertise, a 'feel' for the tool and a measure of confidence or otherwise in the results be established. As with more conventional tools, it is only through often long experience that it is learnt what can and cannot be done. In practice, before undertaking a particular analysis it should always be the policy to tackle a series of simplifications of the original problem in order to further establish confidence in the software as a design tool, and to verify the validity of the abstraction in the particular finite element model selected. As has already been emphasised, the manner in which the model is dissected into finite elements often has a crucial bearing on the final results. For a working guide to these matters the reader is referred to ref. 5.

2.6. FINITE ELEMENT SYSTEMS

There remains the question of which finite element system to use. For many it will be a case of Hobson's choice, of having to use whatever software design tool is available within their particular organisation. Such users may be prone to the conviction that other people's software is better but, like other working conditions, this may be no more than another example of the 'grass is greener on the other side' syndrome. Despite being stuck with a particular system, such users do enjoy the considerable advantage of being able to draw on the body of locally available experience, an advantage not to be lightly taken for granted. Writing as one who had no option but to use a particular system, I have always appreciated and benefitted from the readiness of the company involved to discuss in a full, frank and friendly manner any problems arising both at an individual and a user group level. It has been extremely useful to have the mainframe source code available for my own inspection and modification. Users in the position of being able to choose a software design tool may well find difficulty in evaluating the claims of rival systems. There are many considerations, including those of purchase price, licence agreement, installation and maintenance charges, educational discounts, machine range availability and interface facilities to graphical and other peripheral devices, source code availability, quality of documentation and user courses. If one of these claims is not overriding there remains the suspicion that it is the user response to the sales pitch that decides the issue. However, there are a number of practical steps which may be taken so that an informed decision may be made. In order to ascertain the nature of design tool software currently available and the range of problems addressed, in the area of thin-walled structures a large number of software companies and research organisations were contacted and asked to complete a questionnaire. Appendices 2 and 3 give the results of that survey. Similar, but more extensive and older, surveys are available.[6,7] At the time of writing it appears that there is no completely reliable means for evaluating the various software systems currently on offer. It is even doubtful that such a situation might ever exist, bearing in mind the diversity of problems and computers involved. The NAFEMS organisation[8] has begun to take tentative steps towards establishing standards for software for finite element analysis. Test problems are under construction and the reports of the results of these problems as analysed by various software systems are available to members. Similar reports of commercial systems applied to benchmark problems may also be found.[9]

2.7. CONCLUSIONS

The approach to finite element analysis may be developed using a powerful design tool, the 'software to theory' approach, and contrasts to the 'theory to software' approach which uses software building blocks. Both approaches are used in educational establishments for the teaching of the finite element method and there appears to be no consensus as to which approach is preferable.

However, whichever approach is chosen, care must be exercised. The finite element method depends largely for its accuracy on obtaining a reasonable solution to a system of n-linear equations in n unknowns, n being an integer, and often a very large integer. From a numerical analysis point of view, solving such systems is no straightforward task. Examples abound in the literature[10] of how even the smallest system using standard procedures may yield nonsensical results. There are examples of systems which are so inherently ill conditioned as to defy any computer solution using a finite level of arithmetic accuracy. Therefore, it is essential that all steps are taken, no matter whether the program being used is a simple one-off or a full-blown package, to ensure that only realistic, physically meaningful problems are attempted. In general there is an assumption, well supported by experience, that such problems properly modelled in accordance with the standards laid down in ref. 5 produce well conditioned systems of equations and hence reasonable solutions within tolerable error limits. Commercial programs do have considerable validation procedures but, even so, it is still all too easy for a physically meaningless problem or a badly modelled structure to slip through the net. An elementary data error or a misunderstanding on the part of the user may cause a problem to be solved which may be completely different from the one originally intended. There is the danger that results from the distorted problem may be blithely accepted.

In all computer programs there is an overriding temptation to proceed to a solution; any solution, no matter what happens along the way. No one likes to be seen as admitting failure, or to inconvenience users in the short term by opting out of a calculation halfway through. A momentum builds up which is difficult to resist and so a situation develops whereby any solution is considered preferable to no solution. This is perhaps one area in which many computer programs of a numerical nature, not only finite element programs, are deficient. In addition to providing results it should also be considered essential to provide some measure of confidence, a safety factor. Without such additional information results should be considered inadequate. With the techniques of modern numerical analysis it is perfectly feasible to provide accurate error bounds; yet there is little evidence that programmers are prepared to accept this and, certainly there is little evidence in many of the commercial programs. Sadly, such error estimates are interpreted, by programmer and user alike, as an admission of culpable inaccuracy, a single answer being easier to handle than a range of values. For the time being at least, it seems that engineers will still need to build in their time-honoured safety factors, often more generous than necessary in order to mask computer rounding errors.

REFERENCES

1. Meravides, D. and Turbin, M., Dynamic impact comparisons between experimental and theoretical results obtained with Dyna-3D, *Finite Element News* (1), 1986, 16–19.
2. Strang, G. and Fix, G. J., *An Analysis of the Finite Element Method,* Prentice-Hall, Englewood Cliffs, N.J., 1973.
3. NAG Subroutine Library. For information contact: Numerical Algorithms Group Ltd, NAG Central Office, Mayfield House, 256 Banbury Road, Oxford OX2 7DE, UK.
4. Bennett, J., Creary, L., Englemore, R. and Melosh, R., *A Knowledge-based Consultant for Structural Analysis,* Computer Science Dept., Stanford University, Stanford, California, 1978.
5. Mair, W. M., *Guidelines to Finite Element Practice—NAFEMS,* Department of Trade and Industry National Engineering Laboratory, East Kilbride, Glasgow G75 0QU, UK.
6. Niku-Lari, A. (Ed.), *Structural Analysis Systems,* Vols 1, 2 and 3, Pergamon Press, 1986.
7. Fredriksson, B., and Mackerle, J., *Structural Mechanics Finite Element Computer Programs—Surveys and Availability,* Linkoping Institute of Technology, S-581 83 Linkoping, Sweden.
8. NAFEMS. For information contact: National Agency for Finite Element Methods and Standards, National Engineering Laboratory, East Kilbride, Glasgow G75 0QU, UK.

9. *Finite Element News,* John Robinson and Associates, Great Bidlake Manor, Bridestowe, Okehampton, Devon EX20 4NT, UK.
10. RICE, J. R., *Matrix Computations and Mathematical Software,* McGraw-Hill, New York, 1985.

APPENDIX 1: SOFTWARE BUILDING BLOCKS FOR FINITE ELEMENT ANALYSIS

A1.1. Name of Software: MODULEF

Contact: MODULEF, INRIA-Domaine de Voluceau, Rocquencourt, B.P. 105, 78153 Le Chesnay Cedex, France.
Brief description: General-purpose finite element library of FORTRAN programs developed by the Club Modulef. The club, which was created by INRIA (Institut de Recherche en Informatique et an Automatique) in 1974, brings together universities and industrial companies from several countries, in order to design and implement a library of finite element modules. Existing module capabilities include solution to heat conduction, elasticity and fluid mechanics problems.

A1.2. Name of Software: NAG Finite Element Library

Contact: Numerical Algorithms Group Ltd, NAG Central Office, Mayfield House, 256 Banbury Road, Oxford OX2 7DE, UK.
Brief description: Collection of FORTRAN programs for addressing steady state and time-dependent problems in up to three dimensions. The programs are provided in fully documented source text form and it is envisaged that users will make changes to suit their own requirements. The NAG finite element library was supported by the SERC (Science and Engineering Research Council).

A1.3. Name of Software: SAP

Contact: NISEE/Computer Applications, Davis Hall, University of California, Berkeley, CA 94720, USA.
Brief description: Collection of FORTRAN programs designed to be modified and extended by the user for the static and dynamic analysis of structures.

A1.4. Name of Software: SMUG

Contact: W. Pilkey, Structural Members Users Group, PO Box 3958, Charlottesville, VA 22903, USA.

Brief description: The Structural Members Users Group offers FORTRAN programs for the static, stability, dynamic response and stress analysis of structural members and mechanical elements.

APPENDIX 2: GENERAL-PURPOSE DESIGN TOOLS

For each system a grid of information is supplied. The information is grouped under the three headings namely (I) problem areas, (II) results available to users, and (III) general implementation details. The elements of the grid are numbered in the following manner:

1	2	3	4	5	6	7	8	9	10	11	12	13	14	15	16	17	18	19	20	21	22	23	24	25	26	27
28	29	30	31	32	33	34	35	36																		
37	38	39	40	41	42	43	44	45	46	47	48	49	50													

and contain abbreviated answers to the questions shown below.

(*I*) *Problem areas:* which of the following problem areas are addressed by the software?

(1) Internal/external pressure loading.
(2) Attached load and/or stress carrying members.
(3) Internal/external webbed supports.
(4) Diverse, composite and reinforced materials.
(5) Riveted plates.
(6) Welded construction.
(7) Hydrostatic loading.
(8) Vibrational problems, dynamic stability/instability, centrifugal effects.
(9) Elasticity.
(10) Plasticity.
(11) Variable thickness, variable material properties.
(12) Bending, torsion, compression, shear.
(13) Buckling, post-buckling analysis.
(14) Creep.
(15) Fatigue.
(16) Thermal loading, stresses and buckling.
(17) Uniform/non-uniform pressure pulses.
(18) Impact.
(19) Stress wave propagation.

(20) Acoustic problems.
(21) Aerodynamics, supersonic flow, panel flutter.
(22) Crack propagation, fractures.
(23) Earthquake stresses.
(24) Failure criteria.
(25) Discontinuity stresses.
(26) Radiation effects.
(27) Other areas not included in the above.

(*II*) *Results available to users:* which of the following results are readily available to users of the software?

(28) Stress.
(29) Strain.
(30) Strain rate.
(31) Stress distribution.
(32) Displacements.
(33) Plastic deformation.
(34) Strength—design curve.
(35) Probability of failure.
(36) Others not included in above.

(*III*) *Implementation details:* which of the following apply to the software?

(37) Software attempts to ensure that only realistic, solvable problems are presented.
(38) User informed as to the reliability or otherwise of computed results.
(39) Software modified in the light of user experience.
(40) Wide range of mainframe and smaller computer implementations.
(41) Work-station implementations.
(42) Wide range of graphics devices supported.
(43) Interface to the Graphics Kernel System, GKS.
(44) Interactive facility.
(45) Expert system facility.
(46) Program source code available for user inspection and modification.
(47) Consultancy and advice service.
(48) User courses provided.
(49) User group.
(50) Educational discount.

The information in each grid is coded thus: 'Y' indicates a positive response to a question, 'p', 'possibly', indicates a positive response with some qualification (for example, a solution may be found using a certain ingenuity). In the case of a general implementation question a 'p' might indicate that certain restrictions apply. A 'd' indicates a facility under development. An 'n' implies a negative response.

A2.1. Name of Software: ABAQUS

Y	Y	Y	Y	Y	Y	Y	Y	Y	Y	Y	Y	Y	Y	p	Y	Y	Y	Y	Y	n	Y	Y	Y	p	Y	Y
Y	Y	Y	Y	Y	Y	Y	Y	Y																		
Y	p	Y	Y	Y	Y	n	d	d	p	Y	Y	n	Y													

Contact: Hibbitt, Karlsson and Sorensen, Inc., 100 Medway Street, Providence RI 02906, USA. (UK agent: GE CAE International, York House, Stevenage Road, Hitchin, Herts SG4 9DY, UK.)

A2.2. Name of Software: ALSA—Accurate Large Order Structural Analysis Software

Y	Y	Y	p	Y	Y	Y	Y	Y	n	Y	Y	Y	n	n	n	Y	Y	Y	n	p	p	Y	p	Y	n	p
Y	n	n	Y	Y	n	p	p	n																		
Y	Y	p	Y	Y	n	Y	n	p	p	Y	p	Y	Y													

Contact: T.P.A. Structures, D-605, Post Bag X197, Pretoria, 0001, South Africa.

A2.3. Name of Software: ANSYS

Y	Y	Y	Y	Y	Y	Y	Y	Y	Y	Y	Y	p	Y	Y	Y	Y	Y	Y	n	n	Y	Y	Y	Y	Y	Y
Y	Y	Y	Y	Y	Y	n	n	Y																		
Y	Y	Y	Y	Y	Y	n	Y	n	n	Y	Y	Y	Y													

Contact: Swanson Analysis Systems Inc., Johnson Road, P.O. Box 65, Houston, PA 15342, USA. (UK agent: Structures & Computers Ltd, 1258 London Road, London SW16 4EJ, UK.)

A2.4. Name of Software: ASAS

Y	Y	Y	Y	Y	Y	Y	Y	Y	Y	Y	Y	Y	n	Y	Y	n	Y	n	n	n	Y	Y	Y	n	n	Y
Y	Y	n	Y	Y	Y	n	n	n																		
p	n	Y	Y	Y	Y	n	Y	n	n	Y	Y	Y	Y													

Contact: H. G. Engineering, 260 Lesmill Road, Don Mills, Ontario M3B 2T5, Canada. (UK agent: Atkins Research and Development, Woodcote Grove, Ashley Road, Epsom, Surrey KT18 5BW, UK.)

A2.5. Name of Software: BERSAFE

Y	Y	Y	Y	p	Y	Y	p	Y	Y	Y	Y	p	Y	p	p	Y	Y	n	p	n	Y	Y	p	p	Y	Y
Y	Y	Y	Y	Y	Y	n	n	Y																		
p	p	Y	Y	Y	Y	n	Y	n	Y	Y	Y	p	Y													

Contact: Berkeley Nuclear Laboratories, Berkeley, Gloucestershire GL13 9PB, UK.

A2.6. Name of Software: CASTEM

Y	Y	Y	Y	n	n	Y	Y	Y	Y	Y	Y	Y	Y	Y	Y	Y	Y	Y	Y	n	Y	Y	Y	p	n	n
Y	Y	p	Y	Y	Y	Y	Y	Y																		
n	Y	n	Y	Y	Y	n	p	n	Y	Y	Y	Y	n													

Contact: Informatique Internationale, Agence De Saclay, CEN BP 24, 91190 Gif sur Yvette, France. (UK Agent: SIA Ltd, Engineering Department, Ebury Gate, 23 Lower Belgrave Street, London SW1, UK.)

A2.7. Name of Software: CASTOR

Y	Y	Y	n	Y	Y	Y	Y	Y	Y	Y	Y	p	Y	Y	Y	Y	n	n	n	n	Y	Y	Y	n	n	Y
Y	Y	Y	Y	Y	Y	n	n	Y																		
n	n	n	Y	Y	Y	Y	Y	n	Y	Y	Y	n	p													

Contact: M. Afzali, CETIM, B.P. 67—60304 Senlis Cédex, France. (UK agent: Dr S. Holt, Amazon Computers Ltd, Sunrise Parkway, Milton Keynes MK14 6LQ, UK.)

A2.8. Name of Software: COSMOS

Y	Y	n	p	Y	Y	Y	Y	Y	Y	Y	Y	Y	Y	n	Y	Y	Y	n	n	n	n	Y	Y	Y	Y	Y
Y	Y	Y	Y	Y	Y	n	n	Y																		
Y	n	Y	Y	Y	Y	d	Y	n	Y	n	Y	Y	Y													

Contact: V. I. Weingarten, 1661 Lincoln Boulevard, Suite 100, Santa Monica, CA 90404, USA.

A2.9. Name of Software: EASE2

Y	p	p	n	n	n	n	Y	p	n	n	Y	n	n	n	Y	Y	Y	n	Y	n	n	Y	n	n	n	n
Y	Y	Y	Y	Y	n	Y	n	n																		
Y	Y	Y	Y	n	n	n	p	n	n	Y	Y	n	Y													

Contact: Engineering Analysis Corporation, 24050 Madison Street, Suite 211, Torrance, CA 90505, USA.

A2.10. Name of Software: FEMFAM

Y	Y	n	p	n	Y	Y	p	Y	n	Y	Y	n	n	n	p	n	Y	n	n	n	Y	n	Y	Y	Y	n
Y	Y	Y	Y	Y	n	n	p	n																		
n	p	Y	Y	Y	Y	n	Y	n	Y	Y	p	n	Y													

Contact: Profem GmbH, Salvatorstraße 32, 5100 Aachen, W. Germany.

A2.11. Name of Software: FEMPAC

Y	Y	n	Y	n	n	Y	Y	Y	n	Y	Y	n	n	n	p	p	n	n	n	n	n	n	Y	p	n	n
Y	n	n	p	Y	n	n	n	n																		
Y	p	Y	Y	Y	Y	Y	Y	n	p	Y	Y	n	Y													

Contact: Femprog AB, P.O. Box 26016, S-10041 Stockholm, Sweden.

A2.12. Name of Software: MARC

Y	Y	Y	Y	Y	Y	Y	Y	Y	Y	Y	Y	Y	Y	Y	Y	Y	Y	Y	n	n	Y	Y	Y	Y	Y	Y
Y	Y	Y	Y	Y	Y	n	n	n																		
p	p	Y	Y	Y	Y	n	n	n	p	Y	Y	n	Y													

Contact: MARC Analysis Research Corporation, 260 Sheridan Avenue, Suite 200, Palo Alto, CA 94360, USA. (European agent: MARC-Europe, Bredewater 26, 2715 CA Zoetermeer, The Netherlands.)

A2.13. Name of Software: MEF

Y	Y	Y	Y	p	p	Y	p	Y	Y	p	Y	Y	n	n	Y	n	Y	Y	p	n	Y	n	Y	n	n	n
Y	Y	Y	d	Y	Y	n	n	Y																		
p	n	Y	Y	Y	Y	n	Y	n	Y	Y	Y	Y	Y													

Contact: Jean Francois Cochet, CSI, BP 233—60206 Compiègne Cédex, France.

A2.14. Name of Software: NE-XX

Y	Y	Y	Y	Y	n	Y	p	Y	n	Y	Y	n	n	n	p	n	Y	n	n	n	n	n	n	p	n	Y
Y	n	n	Y	Y	n	n	n	Y																		
p	Y	Y	Y	p	Y	n	Y	n	Y	p	p	n	Y													

Contact: Ivan Nemec, Technical Institute, Dopravoprojekt, 658 30 Brno, Leninova, Czechoslovakia.

A2.15. Name of Software: PAFEC

Y	Y	Y	Y	Y	Y	p	Y	Y	Y	Y	Y	p	Y	n	Y	n	n	n	Y	n	Y	Y	Y	Y	n	Y
Y	Y	n	Y	Y	Y	n	Y	Y																		
Y	Y	Y	Y	Y	Y	Y	Y	p	Y	Y	Y	Y	Y													

Contact: Paul Wheeler, PAFEC Ltd, Strelley Hall, Nottingham NG8 6PE, UK.

A2.16. Name of Software: PUCK-2

Y	Y	n	n	n	n	Y	Y	Y	Y	Y	Y	Y	n	n	Y	Y	n	n	n	n	n	Y	n	n	n	n
Y	n	n	Y	Y	Y	n	n	n																		
n	n	n	Y	n	Y	n	n	n	n	Y	n	n	n													

Contact: CISE S.p.A., via Reggio Emilia, 39-200090 Segrate (Milan), Italy.

A2.17. Name of Software: STRUDL

Y	Y	n	p	n	n	Y	p	Y	p	Y	Y	p	p	n	Y	n	p	n	n	n	n	Y	n	n	n	n
Y	Y	n	Y	Y	Y	n	n	n																		
n	p	Y	Y	n	Y	Y	p	n	Y	Y	Y	Y	Y													

Contact: Messerschmitt–Bolkow–Blohm GmbH, PO Box 801160, 8000 München 80, W. Germany.

A2.18. Name of Software: WECAN

Y	Y	Y	Y	n	n	Y	Y	Y	Y	Y	Y	p	Y	Y	Y	Y	n	Y	Y	n	Y	p	n	n	Y	n
Y	Y	Y	Y	Y	Y	n	n	n																		
p	Y	Y	Y	Y	Y	p	Y	n	p	n	p	Y	Y													

Contact: Engineering Service Bureau, Advanced Mechanics Services, Box 355, Pittsburgh, PA 15230, USA.

Many of the software systems already mentioned are following the trend towards complete computer-aided design systems by providing extensive facilities for the modelling and display of analysed structures. The following systems are primarily regarded as modelling tools which, although having in some cases a local capability for finite element analysis, do in addition provide an interface to one or more of the systems already mentioned. Implementations are available on a wide range of mainframes and work-stations.

A2.19. Name of Software: FEGS

Contact: FEGS Ltd, Oakington, Cambridge, CB4 5BA, UK.

A2.20. Name of Software: GIFTS

Contact: N.A.P. Engineering Microsoftware Ltd, 22a Ryders Terrace, St. Johns Wood, London NW8 0EE, UK.

A2.21. Name of Software: PATRAN II

Contact: PDA Engineering International, 1833 East Alton Avenue, Irvine, CA 92714-4902, USA. (UK agent: PDA Engineering International, Winton House, Winton Square, Basingstoke, Hants RG21 1EN, UK.)

APPENDIX 3: SPECIFIC DESIGN TOOLS

The following design tools are of a more restricted nature, being limited to specific problem areas and/or limited to implementation on a restricted number of computer ranges or computer operating systems. An asterisk (*) denotes that the source code is readily available for user inspection and modification.

A3.1. Name of Software: ARS–ASE, TEKTON–ASE
Contact: Dr C. Katz, Roemerweg 1, D-8138 Andechs, W. Germany.
Brief description: Programs for 3-D analysis of thin-walled structures. Comprehensive system for civil engineering applications.
Implementation: MS-DOS.

A3.2. Name of Software: AXISYMMETRIC*
Contact: Dr C. T. F. Ross, 6 Hurstville Drive, Waterlooville, Portsmouth PO7 7NB, UK.
Brief description: Suite of programs written in BASIC for the static, dynamic and stability analysis of thin-walled and thick-walled axisymmetric structures.
Implementation: Most microcomputers, minis and mainframes.

A3.3. Name of Software: BEOS*
Contact: Dr Ing. B. Geier, DFVLR—Institut für Strukturmechanik, Postfach 3267, D-3300 Braunschweig, W. Germany.
Brief description: Program for calculation of buckling loads and natural vibrations of eccentrically orthotropic shells. Layered and stiffened shells are included. Particularly useful for aerospace structural components.
Implementation: OS/MVS, VAX/VMS.

A3.4. Name of Software: BOSOR4*
Contact: David Bushnell, 92-50/255 Lockheed Applied Mechanics, 3251 Hanover Street, Palo Alto, CA 94304, USA.
Brief description: Program for stress, buckling and modal vibration analysis of ring-stiffened, branched shells of revolution loaded either axisymmetrically or non-symmetrically. Complex elastic wall construction permitted. Buckling and/or stress response to harmonic or random base excitation.
Implementation: VAX, IBM, PRIME.

A3.5. Name of Software: BOSOR5*
Contact: As above.
Brief description: Program for performing axisymmetric collapse and

non-symmetric bifurcation buckling, including elastic–plastic material behaviour and creep.
Other details: as for BOSOR4.

A3.6. Name of Software: ESA
Contact: SCIA, Steenweg 108, 3912 Herk-de-stad, Belgium. (UK agent: Eclipse Associates Ltd, Lovat Bank, Silver Street, Newport Pagnell, Milton Keynes MK16 0EJ, UK.)
Brief description: Large collection of programs for finite element analysis including static elastic analysis, dynamic and stress analysis of beam, plate and shell structures. Extensive graphical display facilities.
Implementation: Programs developed on Wang mainframes and work-stations.

A3.7. Name of Software: FESDEC
Contact: Dr A. Firmin, H. G. Engineering, 260 Lesmill Road, Don Mills, Ontario M3B 2T5, Canada.
Brief description: General-purpose finite element system written specially for Hewlett-Packard desktop computers. Large element library providing facility for analysing all classes of structures.
Implementation: Hewlett-Packard.

A3.8. Name of Software: FLASH
Contact: Dr D. Pfaffinger, P & W Engineering, Forchstrasse 21, PO Box 231, CH-8032 Zurich, Switzerland.
Brief description: General-purpose program for the analysis of beam structures, membranes, folded plates and shells. Program is suitable for the analysis of concrete slabs, bridges, steel constructions and machine parts.
Implementation: Many mainframes and IBM PC-AT work-station.

A3.9. Name of Software: KYOKAI*
Contact: Dr Kazuei Onishi, Applied Mathematics Department, Fukuoka University, Fukuoka 814-01, Japan.
Brief description: Menu-driven command system for the solution of two dimensional potential and thermo-elasticity problems in zoned inhomogeneous materials.
Implementation: Any IBM-compatible system.

A3.10. Name of Software: MacFEM
Contact: Numerica S.A.R.L., 23 Bd. de Brandenbourg BP 215, Ivry/Seine 94203, France.
Brief description: Collection of programs for structural analysis.
Implementation: Microcomputers in general and the Macintosh in particular.

A3.11. Name of Software: NOLIN
Contact: Dr Ing. B. Geier, DFVLR—Institut für Strukturmechanik, Postfach 3267, D-3300 Braunschweig, W. Germany.
Brief description: Program for investigating the non-linear behaviour of plates and curved panels, including non-linear pre-buckling and post-buckling.
Implementation: OS/MVS, CRAY-COS.

A3.12. Name of Software: PANDA*
Contact: As for BOSOR4.
Brief description: Program for the optimisation with respect to weight of cylindrical panels or cylindrical shells with discrete or smeared stringers and/or rings, subjected to simultaneous uniform in-plane axial compression, hoop compression and shear. Facilities for plasticity and general and local buckling analysis.
Implementation: VAX, but should convert easily to other computers.

A3.13. Name of Software: PSTAR*
Contact: Dr Zsolt Revesz, PO Box 1126, CH-5401 Baden, Switzerland.
Brief description: Series of computer programs for the linear–elastic analysis of three-dimensional piping systems with graphic presentation of selected results.
Implementation: MS-DOS, PRIMOS, CYBER-NOS.

A3.14. Name of Software: ROBOT*
Contact: I. Paczelt, Department of Mathematics, Technical University, H-3515 Miskolc Egyetemvaros, Hungary.
Brief description: Computer program for linear–elastic analysis of axisymmetrical structures subject to axisymmetrical loading. A great variety of structures may be analysed, including segmented, branched

shells of revolution with various meridional properties. Temperature-dependent material property may range from homogeneous isotropic to non-homogeneous orthotropic.
Implementation: CDC, VIDETON.

A3.15. Name of Software: SAFE-SHELL* (Stress Analysis of Thin Shells), STRAP* (Structural Analysis Package), SLADE* (Transient Dynamic Response of Elastic Shells), and Many Others

Contact: Margaret Butler, National Energy Software Center, Argonne National Laboratory, Argonne, IL 60439, USA.
Brief description: The centre maintains a large collection of finite element software to cover a wide range of problems, too numerous to detail here.
Implementation: Mainly FORTRAN programs written for mainframes.

A3.16. Name of Software: STAGSC*

Contact: John Gibson, COSMIC, Suite 112, Barrow Hall, University of Georgia, Athens, GA 30602, USA.
Brief description: Computer code primarily intended for the analysis of structural shells. Originally developed to analyse the non-linear behaviour of cylindrical shells with cutouts. Subsequently, scope extended to include spring and beam elements. Options for analysis of linear and non-linear static stress, stability, vibrations and transient response.
Implementation: VAX, CDC, CRAY, IBM.

A3.17. Name of Software: STRESS

Contact: R. G. Reed, Computational Mechanics International, Ashurst Lodge, Ashurst, Southampton SO4 2AA, UK.
Brief description: Microcomputer programs to perform structural analysis of two- and three-dimensional structures composed of slender members.
Implementation: MS-DOS, CP/M 80.

A3.18. Name of Software: TSTAR*

Contact: As for PSTAR.
Brief description: Series of computer programs for the linear–elastic

and elasto-plastic stress and thermal stress analysis of T junctions, and for graphic presentation of selected results.
Implementation: MS-DOS, PRIMOS, UNIX.

A3.19. Name of Software: ZERO-3
Contact: CISE S.p.A., via Reggio Emilia, 39-200090 Segrate (Milan), Italy.
Brief description: Program for the static and dynamic analysis of axisymmetric structures with non-symmetrical loads, taking into account dynamic fluid structure interaction effects.
Implementation: VM/CMS, VAX.

Chapter 3

Detecting and Avoiding Numerical Difficulties

ED AKIN

Department of Mechanical Engineering and Materials Science, Rice University, Houston, Texas, USA

3.1. INTRODUCTION

The verification of data for finite element models of thin-walled structures and the avoidance of numerical difficulties is of increasing importance. A number of verification checks are possible and will be discussed. They are based on both theoretical and empirical studies. Most of the checks can be automated to enhance the reliability of its finite element codes.

Finite element analysis techniques[1–4,17] require extensive amounts of input data. Similarly, they usually generate extensive amounts of data output. Competent engineering analysis requires that both of these data sets be acceptable in length. The goal of this section is to outline several program verification procedures that can be utilized by the analyst to assure the correctness of a model or solution.

The importance of procedures for verifying the model correctness is often overlooked in finite element texts. Exceptions are the works of Cook,[2] Irons[4] and elementary discussions in ref. 5. They provide brief chapters that include several useful observations on improper or ill-conditioned data or elements. However, there are several other useful verification procedures. The initial emphasis here will be on the checks that can be utilized to verify input data. These can be generally broken into five major areas: basic data, validation, element geometry, element distortion and consistency. Mesh validation, certain special features of thin-walled structures, load validation, and errors in basic equations will be considered, followed by the validation of output data.

3.2. WHERE ERRORS CAN OCCUR

3.2.1. Data Validation

The data validation stage involves checking the input data for features that could lead to unproductive, incorrect, or expensive computer usage. For example, a statics analysis should check that a load case has been defined and warn if loads are applied to a restrained degree of freedom. The syntax of the data should be scanned to check for typing errors. Input data should also be checked to verify that omitted items are defaulted correctly. The most modern codes have a data dictionary (DD) system that avoids these problems by defining defaults. If an entry cannot be defaulted and has not been entered, then an error will be given. Also, it is important that input data modules should be examined to ensure that an entry does not conflict with data supplied in other modules. These types of facilities can provide considerable savings in time and money.

Programs using modern computer science techniques employ a data base management system (DBMS). One advantage of a DBMS is that it allows the user to delete or append additional data without repeating the complete input phase. These changes or corrections can often be done on an interactive basis. Also, since data validation can be expensive, the user should be given the option to skip such checks on files that have been previously checked. The text by Irons[4] includes programs, with numerous comments, for carrying out extensive data validation checks.

One obvious way of checking data is to allow extensive plotting options. Mesh plots should include the usual solid-line plots plus other options that would identify holes or slits in the region due to omitted or incorrect data. Such plots include exploded views of all elements, boundary-line plots, or dashed-line plots for shared interior lines. Hidden-line plots are helpful and can often be obtained by interfacing with interactive codes, such as the public domain MOVIE.BYU program or commercial codes like PATRAN. Other user controls, such as specified viewing points, should be available.

Graphical verification of the data is enhanced if extra information can be included with the mesh plots. The following information should be available for such displays: node point locations, node numbers, element numbers, material numbers, axis sets, element wavefront or bandwidth, restraints, degree-of-freedom numbers on arrows, and load vectors. Since this amount of data can clutter a plot, the analyst should be able to specify plots by group numbers, or element number range, or by some interior or exterior windowing commands.

There is an increasing trend towards the use of interactive computing in computer-aided design. It is therefore quite useful if portions of the verification checks can be conducted interactively. These interactive graphics programs usually allow for the following features: addition or deletion of nodes or elements, display and change topology, display and change co-ordinates, plot mesh boundary, display node or element labels, rotate the mesh, window or zoom in on regions, show undeformed or deformed mesh, etc. A few programs also offer digitizing options to define nodes and elements, and allow for interactive mesh generation.

The availability of codes designed to assist in data verification is increasing. Features such as free format input, DD, DBMS, digitizing and interactive graphics are most useful in reducing errors and increasing productivity. The major finite element codes, such as ANSYS, MARC, PAFEC, PATRAN, SUPERB and others, offer one or more of these features. Some offer all of these capabilities. Program comparisons[6,7,15] list specific features and differences in capabilities in commonly used software. However, it will always be important for the analyst to use good engineering judgement when building finite element models, and not rely on programs to catch bad procedures.

3.2.2. Element Validation

The successful application of finite element analysis should always include a validation of the element to be used and its implementation in a specific computer program. That is especially true for applications involving thin-walled structures. Usually, the elements utilized in other problem classes, such as plane stress, are very well understood and tested. However, shells, thin-walled beams, and thin-walled structures can be difficult to model and the elements used for their analysis are usually more prone to numerical difficulties.[18] Therefore, one should subject elements to be used to a series of element validation tests. Two of the most common and important tests are the patch test introduced by Irons,[4,5,8] and the single-element tests proposed by Robinson.[9,10] The single-element tests generally show the effects of element geometrical parameters such as convexity, aspect ratio, skewness, taper, and out-of-plane warpage. It is most commonly utilized to test for a sensitivity to element aspect ratio. Thin-walled structural elements have historically been found to be likely to develop numerical difficulties when one of these geometrical parameters varies over a wide range. The single-element test[9,11] usually consists of taking a single element in rectangular, triangular, or line form, considering it as a complete structure and then investigating its behavior for various load or boundary conditions as a geometrical parameter is varied. An analytical solution is usually available for such a structure. Such a test is very helpful in understanding the load-carrying capabilities of the element. It also helps to demonstrate its sensitivity to parameters such as aspect ratio. Application of the single-element test to several common plate and shell elements[11] used in thin-walled structures shows that a large number of them have very significant errors when geometrical parameters are distorted. Deflection errors of 50–100% are not uncommon in such tests. This is especially true for the pure twist load states which are often encountered in the application of thin-walled structures. Some typical numerical results will be considered in a later section.

The patch test has been proven to be a valid convergence test.[4] It was developed from physical intuition and later written in mathematical forms. The basic concept is fairly simple. Imagine what happens as one introduces a very large, almost infinite, number of elements. Clearly, they would become very small in size. If we think of the quantities being integrated to form the element matrices we can make an observation about how the solution would behave in this limit. The integrand, such as the strain energy, contains derivative terms that would become constant as the element size shrinks toward zero. Thus, to be valid in the limit, the element formulation must be able to yield the correct results in that state. That is, to be assured of convergence one must be able to exactly satisfy the state where the derivatives, in the governing integral statement, take on constant or zero values. This condition can be stated as a mathematical test or as a simple numerical test. The latter option is what we want here. The patch test provides a simple numerical way for a user to test an element, or complete computer program, to verify that it behaves as it should.

We define a patch of elements to be a mesh where at least one node is completely surrounded by elements. Any node of this type is referred to as an interior node. The other nodes are referred to as exterior or perimeter nodes. We will compute the dependent variable at all interior nodes. The derivatives of the dependent variable will be computed in each element. The perimeter nodes are utilized to introduce the boundary conditions and/or loads required by the test. Assume that the governing integral statement has derivatives of order n. We would like to find boundary conditions that would make those derivatives constant. This can be done by selecting an arbitrary nth order polynomial function of the global co-ordinates to describe the dependent variable in the global space that is covered by the patch mesh. Clearly, the nth order derivatives of such a function would be constant as desired. The assumed polynomial is used to define the boundary conditions on the perimeter nodes of the patch mesh.

This is done by substituting the input co-ordinates at the perimeter nodes into the assumed function and computing the required value of the dependent variable at each such node. Once all of the perimeter boundary conditions are known the solution can be numerically executed. The resulting values of the dependent variable are computed at each interior node. To pass the patch test these computed internal values must agree with the value found when their internal nodal co-ordinates are substituted into the assumed global polynomial. However, the real test is that when each element is checked the calculated nth order derivatives must agree with the arbitrarily assumed values used to generate the global function. If an element does not satisfy this test it should not be used. The patch test can also be used for other purposes. For example, the analyst may wish to distort the element shape and/or change the numerical integration rule to see what effect that has on the numerical accuracy of the patch test.

As a simple elementary example of an analytic solution of the patch test, consider the bar element. The smallest possible patch is one with two line elements. Such a patch has two exterior nodes and one interior node. For simplicity, let the lengths of the two elements be equal and have a value of L.

The governing integral statement contains only the first derivative of u. Thus an arbitrary linear function can be selected for the patch test since it would have a constant first derivative. Therefore select $u(x) = a + bx$ for $0 \leqslant x \leqslant 2L$, where a and b are arbitrary constants. Assembling the two-element patch gives

$$\frac{AE}{L}\begin{bmatrix} 1 & -1 & 0 \\ -1 & (1+1) & -1 \\ 0 & -1 & 1 \end{bmatrix}\begin{Bmatrix} u_1 \\ u_2 \\ u_3 \end{Bmatrix} = \begin{Bmatrix} F_1 \\ 0 \\ F_3 \end{Bmatrix}$$

where F_1 and F_3 are the unknown reactions associated with the prescribed external displacements. These two exterior patch boundary conditions are obtained by substituting their nodal co-ordinates in the assumed patch solution:

$$u_1 = u(x_1) = a + b(0) = a$$
$$u_3 = u(x_2) = a + b(2L) = a + 2bL$$

Modifying the assembled equations to include the patch boundary conditions gives

$$\frac{AE}{L}\begin{bmatrix} 0 & -1 & 0 \\ 0 & 2 & 0 \\ 0 & -1 & 0 \end{bmatrix}\begin{Bmatrix} u_1 = a \\ u_2 \\ u_3 = a + 2bL \end{Bmatrix}$$
$$= \begin{Bmatrix} F_1 \\ 0 \\ F_3 \end{Bmatrix} - \frac{aAE}{L}\begin{Bmatrix} 1 \\ -1 \\ 0 \end{Bmatrix} - \frac{(a + bL)AE}{L}\begin{Bmatrix} 0 \\ -1 \\ 1 \end{Bmatrix}$$

Retaining the independent second equation gives the displacement relation

$$\frac{2AE}{L}u_2 = 0 + \frac{aAE}{L} + \frac{(a + 2bL)AE}{L}$$

Thus the internal patch displacement is

$$u_2 = (2a + 2bL)/2 = (a + bL)$$

The value required by the patch test is

$$u(x_2) = (a + bx_2) = (a + bL)$$

This agrees with the computed solution, as required by a valid element. Recall that the element strains are defined as follows:

$$e = 1: \quad \varepsilon = (u_2 - u_1)/L = [(a + bL) - a]/L = b$$
$$e = 2: \quad \varepsilon = (u_3 - u_2)/L = [(a + 2bL) - (a + bL)]/L = b$$

Thus all element derivatives are constant. However, these constants must agree with the constant assumed in the patch. That value is

$$\varepsilon = \mathrm{d}u/\mathrm{d}x = \mathrm{d}(a + bx)/\mathrm{d}x = b$$

Therefore, the patch test is completely satisfied. At times one also wishes to compute the reactions, i.e. F_1 and F_3. To check for possible rank deficiency in the element formulation one should repeat the test with only enough displacements prescribed to prevent rigid body motion. Then the other outer perimeter nodes are loaded with the reactions found in the previous patch test. In the above example, substituting u_1 and u_2 into the previously discarded first equation yields the reaction $F_1 = -bAE$. Likewise, the third equation gives $F_3 = -F_1$, as expected. Thus the above test could be repeated by prescribing u_1 and F_3, or F_1 and u_3. The same results should be obtained in each case.

A major advantage of the patch test is that it can be carried out numerically. In the above case, the constants a and b could have been assigned arbitrary values. Inputting the required numerical values of A, E and L would give a complete numerical description that could be tested in a standard program. Such a procedure also verifies that the computer program satisfies certain minimum requirements.

A problem with some elements is that they can pass the patch test for a uniform mesh but fail when an arbitrary irregular mesh is employed. Thus, as a general rule, one should try to avoid conducting the test with a regular mesh such as the one given in the above example. It would have been wiser to use unequal element lengths such as L and αL, where α is an arbitrary constant. The linear bar element should pass the test for any scaling ratio, α.

3.2.3. Mesh Validation

Several aspects of the problem geometry can be subjected to reasonable checks. Some of the errors in geometry may not be apparent at first observation. For example, consider the use of isoparametric elements used to match curved boundaries or surfaces. Often, quadratic isoparametric elements are utilized to approximate circular arcs. This involves errors in modeling the radius, arc length and tangent angles. Many analysts have no feel for the errors made in selecting their original models. A knowledge of these parameters can thus provide guidance to better input data. Consider a circular arc subtending an angle of 2θ that is approximated by a quadratic isoparametric curve using nodes at each end and the center. This is illustrated in Figs 3.1(a) and 3.1(b). From geometric investigations[12] of these two curves, one can derive the error in the radius, Δ, and in the tangent angles, β. These errors are tabulated for various θ in Table 3.1. Figure 3.1(b) shows how these geometric errors increase rapidly with the angle θ. These errors are not only important in individual

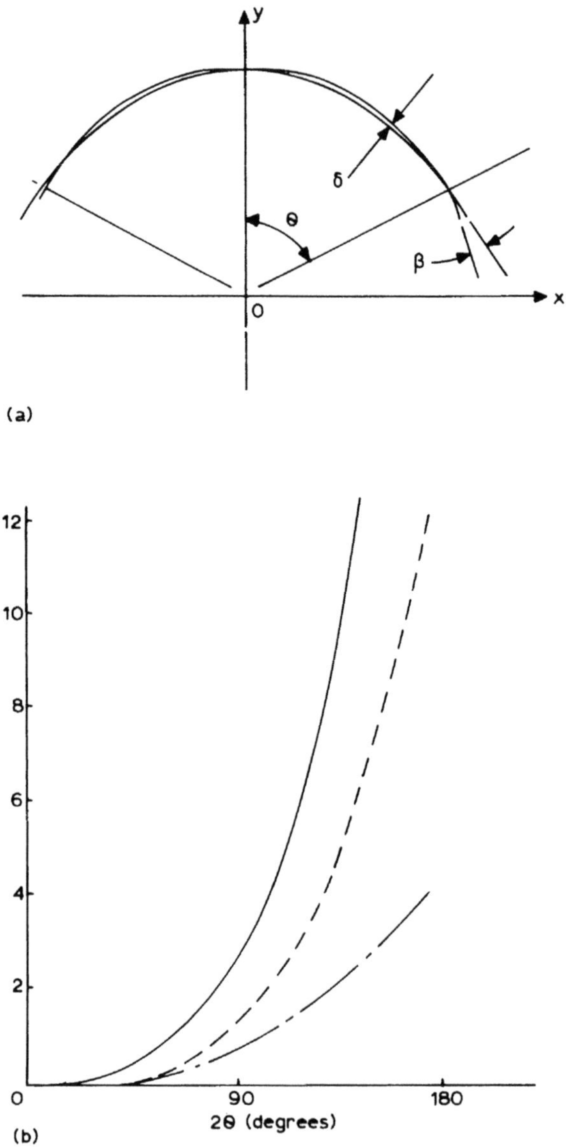

FIG. 3.1. Typical isoparametric geometry errors. ———, β (degrees). ----, $\Delta/R(\%)$. —·—, $\Delta/R\theta(\%)$. $\Delta = \delta_{max}$.

TABLE 3.1
Discrepancies Between Circle and Parabola

Angle 2θ *(degrees)*	*Radius* $\Delta/R(\%)$	$\Delta/(R\theta)(\%)$	*Tangent* β *(degrees)*
20	0·003	0·008	0·03
40	0·045	0·065	0·21
60	0·220	0·210	0·70
80	0·690	0·490	1·71
100	1·610	0·920	3·43
120	3·180	1·520	6·16
140	5·570	2·280	10·24
160	8·930	3·200	16·15
180	13·400	4·260	24·59

elements; they are also important because many codes use isoparametric mesh generation options. Such options tend to use large blocks, and thus large θ, to approximate the original curved geometry. Thus the generated points may not follow the true circle as accurately as possible. Some codes, such as PAFEC, assume that circular arcs are intended as input and make corrections for these geometric errors. Being aware of these possible errors allows a user to more accurately verify data that is to be expanded by mesh generation programs. Later comments on element distortion effects can also pertain to isoparametric mesh generation options.

Other element geometry checks are fairly obvious. These include checks for negative node numbers and warnings for nodes or topology entries that are repeated. A feature of special concern is that of element distortion. Several warning and error conditions can be given to validate various element types.

Another common structure is the thin-walled beam.[3,13,14] This requires the input or calculation of several additional geometrical parameters for inclusion in the element matrices. Usually, a beam in space will have two degrees of freedom to account for axial loads, four degrees of freedom for transverse bending effects in the local y-plane, four for bending in the local z-plane, and finally two for torsion about the local x-axis. However, Kawai[14] and Hughes[3] have shown that for thin-walled beams the two usual degrees of freedom for the St. Venant's torsional terms, GJ, are not sufficient. One must use at least four degrees of freedom for the torsion of the thin-walled beam. Thus the usual 12-degree-of-freedom space beam becomes a 14-degree-of-freedom space beam if warping is restrained. The torsional moment relation is then

$$M_x = GJ\frac{d\theta}{dx} - EI_\omega\frac{d^3\theta}{dx^3}$$

where θ is the angle of twist, and the new second term introduces the warping torsional inertia, I_ω, of the cross-section. The geometric parameters such as the area (for axial loads), the center of twist, the warping function, the warping torsional inertia, and others depend on several line integrals around the thin section plus logic as to whether it is an open or closed section. These parameters are very important to the thin-walled element, but they are difficult to verify. One may wish to develop a pre-processor for the purpose of describing a cross-section and automating these geometric calculations.

It is rare to find such a thin-walled beam torsion element included in programs other than those designed for the structural analysis of ships. One will more often have available only standard space beams to connect to a general thin-walled structure. Only a few validation checks are available for such beams. Most deal with the sensitivity of the matrices for transforming from the local principal axes of the beam to the global axes. The relative orientation of these two axes are commonly defined by one of three options. The so-called axis method uses one of the global axes of the structure and the local x-axis of the beam. They define the plane containing the principal bending (local x–y) plane of the member. The vector cross-products necessary to form the transformation matrix will lose accuracy if the angle between the two axes is less than 5°. At that point the validation checks should issue a warning message. If the angle drops to 1°, or less, then a fatal error should be denoted. Of course, if the angle goes to zero the principal plane is non-unique. The other methods have similar angle limits. The node method uses a third reference node. It defines the principal plane to be the one containing the local x-axis and the third node. In that case the angles are measured from the line connecting the first node and the reference node to the local x-axis (the line connecting nodes 1 and 2). Obviously, the reference node should not be one of the beam nodes, nor can it lie on the same line (a common error). If the beam is curved then the reference node is usually also utilized as the center of curvature of the beam. In that case curved beam theory and empirical studies suggest that a warning should be issued if the angle subtended at the center of curvature (reference node) is greater than 45°. Since circular beams are assumed when a two-node topology is supplied, one should also verify that the distances from the topology nodes to the center of curvature are within about 1% of each other.

The angular orientation method is used in several codes. It uses a fixed global plane (say, x–y) and an angle of rotation about the local beam x-axis. The angle specified is the one that would rotate the line of the local y-axis so that it would be parallel to the selected global plane. Obviously, this approach can not be used if the local beam x-axis is normal to the given global plane (parallel to the global z-axis). The above angle validations are applied to the angle between the beam x-axis and the normal to the global plane.

3.2.4. Element Distortion

The effects of distorting various types of elements can be serious and most codes do not adequately validate data in this respect. As an example, consider a quadratic isoparametric line element. As shown in Fig. 3.2, let the three nodes be located in physical (x) space at points 0, ah, and h where h is the element length and $0 \leqslant a \leqslant 1$ is a location constant. The element is defined in a local unit space where $0 \leqslant s \leqslant 1$. The relation between x and s is easily shown to be

$$x(s) = h(4a - 1)s + h(2 - 4a)s^2$$

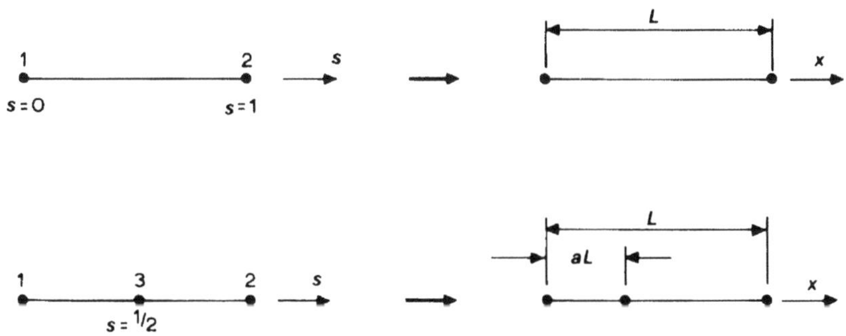

FIG. 3.2. Constant Jacobian and distorted elements.

The two co-ordinates have derivatives related by

$$\frac{\partial x}{\partial s} = h(4a-1) + 4h(1-2a)s$$

The Jacobian of the transformation, J, is the inverse relation; that is, $J = \partial s/\partial x$. The integrals required to evaluate the element matrices utilize this Jacobian. The mathematical principles require that J be positive definite. Distortion of the elements can cause J to go to zero or become negative. This possibility is easily seen in the present 1-D example. If one locates the interior $(s = \frac{1}{2})$ node at the standard midpoint position, then $a = \frac{1}{2}$ so that $\partial x/\partial s = h$ and J is constant throughout the element. Such an element is generally well formulated. However, if the interior node is distorted to any other position, the Jacobian will not be constant and the accuracy of the element may suffer. Generally, there will be points where $\partial x/\partial s$ goes to zero, so the stiffness becomes singular due to division by zero.

For slightly distorted elements, say $0{\cdot}4 \ll a \ll 0{\cdot}6$, the singular points lie outside the element domain. As the distortion increases the singularities move to the element boundary, e.g. $a = \frac{1}{4}$ or $a = \frac{3}{4}$. Eventually, the distortions cause singularities of J inside the element. Such situations cause poor stiffness matrices and very bad stress estimates.

The effects of distortions of two- or three-dimensional elements are similar. For example, the edge of a quadratic element may have the non-corner node displaced in a similar way or it may be moved normal to the line between the corners. Similar analytic singularities can be developed for such elements. However, the presence of singularities due to element distortions can easily be checked by numerical experiments. Several such analytic and numerical studies have led to useful criteria to check the element geometry for undesirable effects.

For example, Fig. 3.3 shows a typical edge of a two- or three-dimensional quadratic element. Let L be the cord length, D the normal displacement of the mid-side node, and α the angle between the corner tangent and the cord line. The criteria show:[8]

warning range: $\frac{1}{7} < D/L < \frac{1}{3}$, $\alpha \leqslant 30°$

error range: $\frac{1}{3} > D/L$, $\alpha \geqslant 53°$

These values are obtained when only one edge is considered. If more

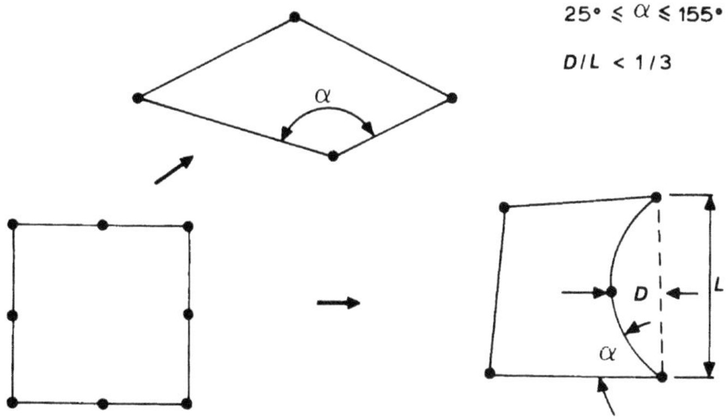

FIG. 3.3. Distorted quadrilaterals.

than one edge of a single element causes a warning state then the warnings should be considered more serious.

Other parameters influence the seriousness of element distortion. Let R be a measure of the aspect ratio. That is, R is the ratio of the longest side to the shortest side. Let the minimum and maximum angles between corner cord lines be denoted by θ and γ, respectively. Define H, the lack of flatness, to be the distance of a fourth node from the plane of the first three divided by the maximum side length. Then the following guidelines in Table 3.2 should be considered when validating geometric data for membrane or solid elements.

TABLE 3.2
Geometric Criteria for Two- and Three-dimensional Elements

Shape	*Warning state*	*Error state*
Triangle	$5 < R < 15$	$R > 15$
	$15° < \theta < 30°$	$\theta \leq 15°$
	$150° < \gamma < 165°$	$\gamma \geq 165°$
Quadrilateral	$5 < R < 15$	$R \geq 15$
	$25° < \theta < 45°$	$\theta \leq 25°$
	$135° < \gamma < 155°$	$\gamma \geq 155°$
	$10^{-5} < H < 10^{-2}$	$H \geq 10^{-2}$

Solids: The above limits on R, θ and γ are checked on each face.

3.3. OUTPUT VALIDATION

3.3.1. Verification Checks

The major emphasis of verification checks is placed on input data. However, it is wise to summarize output checks that the analyst should employ. The type of check depends upon the application involved. The most common applications are stress analysis, structural dynamics and heat transfer. Once again, one traditionally conducts preliminary checks with graphical output. The graphical output that should be available include: deformed shapes, static, mode, or selected time steps, contours of stresses or temperatures, stress-vector plots, failure criterion values, etc. Another useful feature is the ability to select paths through the mesh and plot the variation of displacements, stress or temperatures along such paths. A lack of smoothness in such plots would usually be an indication that the mesh should be finer. Stress plots should use extrapolations from the Barlow points[18] for maximum accuracy.

In time or frequency-dependent problems one would want to plot the time-history or frequency response of the primary variables. In structural dynamics or modes and frequencies calculations, the selection of the master dynamic degree of freedom is very important. Some general guidelines for their selection are available.[2,12] Extensive empirical studies suggest that algorithms can automatically make the selection and often obtain more accuracy than an experienced user. Regardless of how any masters are selected, their locations should be plotted on the mesh so that the analyst can verify the reasonableness for the problem under study.

The validation of finite element input and output data should be a part of every engineering study. While many of the proposed checks are relatively simple, the analyst should keep a mental checklist and verify every analysis. However, the important checks on element distortions should be available within the program, because the number of checks involved is large unless the mesh is trivial.

The output from the program should also include generated data that would be of use in validating the input and for use with other output. For example, a list of elements attached to each node is important, and can be of use in searching the stress file for results near a particular point. If one requests stress averaging at a node then the list is needed for that purpose. It also gives, as a by-product, the list of nodes that are not connected to the mesh. Such lists are usually a part of any 'pre-front' calculations and will therefore be needed in frontal solvers.

The applied pressure, specified temperature or given restraint data are often expanded by mesh generation programs. The supplementation lists created for these data should be printed in the output and checked by the program.

Another output that is useful in data validation is the list of mass properties of the system. Several codes provide the mass, center of mass, mass moments of inertia, and principal moments of inertia for all element groups and for the total mesh. This can give a useful commonsense check of the model, plus providing data for material costs, etc. This data should be calculated even for static problems.

In the analysis of thin-walled structures it is not unusual for flat finite elements used for shells to be assembled in a plane at a node. This causes the normal rotational degree of freedom to have no stiffness and makes the system singular. Such a condition should be detected and corrected. A summary of such corrections should also be listed in the output.

The printed output should provide a summary of the maximum displacements and stresses (or temperatures) found in the analysis. When possible, bar charts should be printed with nodal or element output to help the analyst in spotting the locations of maximum values. Likewise, graphs of values versus node number or co-ordinate location can be of great value. These output items also help to reduce the time required to produce analysis reports.

If significant validation checks are not satisfied then the program should terminate the solution for data corrections and later restart. The extensive use of batch or interactive graphics should be utilized to speed the verification process. These graphics options also have the added benefit of simplifying the preparation of the final analysis reports.

3.4. ERROR ASSESSMENT WITH EXAMPLES

The error assessment of thin-walled structures will often involve thin plates or shells. When utilizing planar quadrilateral plates to analyze flat, cylindrical or conical surfaces, one may unintentionally generate warping effects that are very undesirable. The lack of flatness is usually the result of small geometric errors due to co-ordinate transformations, too few significant figures in the data, or the isoparametric generation errors mentioned earlier. The warping caused by topology errors is usually found in graphical checks of the mesh. The previously mentioned validation checks of flatness should catch other errors. Most flat quadrilaterals are very sensitive to any out-of-plane warping of the geometry. Such warping changes the mapping of a two-dimensional parent to a three-dimensional surface. That means that one must invoke differential geometry to correctly formulate the element. For example, the differential surface area is usually the product of the determinant of the Jacobian and the differential changes in the parent coordinates. When the surface is warped one must replace the determinant of the Jacobian with the square root of the determinant of the first fundamental magnitudes of the curvilinear surface.[1] Thus strains, integrals and stiffnesses quickly develop significant errors, even though a plot may look as if the element is flat.

Out-of-plane warping can result in significant local forces falsely implying constraints on the displacements. This causes errors in the displacements and significant errors in the stresses. A number of procedures can be used to assess the existence of such difficulties in the analysis. One useful procedure is to invoke a rigid body check. Let **K** denote the system stiffness matrix, **F** the generalized nodal forces, and **R** a set of displacement components that represent a rigid body motion. For example, **R** may correspond to a rigid body translation in the x direction. Then **R** would be zero except for constants, say c, associated with each x displacement degree of freedom. We know that such a rigid body displacement should cause no external nodal forces and no internal stresses. Thus, before solving the assembled equations, we execute a rigid body check:

$$\mathbf{KR} = \mathbf{F} \stackrel{?}{=} \mathbf{0}$$

This should identify local errors associated with poor elements or incorrect constraints. Since we generally have three rigid body translations and three rigid body rotations, the array **R** usually contains six columns. Thus the above result should be a null rectangular array. If we applied this test to **K** previously assembled for our patch test we would find $\mathbf{R}^{\mathrm{T}} = [c\ c\ c]$, and

$$\frac{EA}{L}\begin{bmatrix} 1 & -1 & 0 \\ -1 & 2 & -1 \\ 0 & -1 & 1 \end{bmatrix}\begin{Bmatrix} c \\ c \\ c \end{Bmatrix} = \begin{Bmatrix} 0 \\ 0 \\ 0 \end{Bmatrix}$$

as expected. If we had run the distorted patch test, then the left-hand side would be

$$\frac{EA}{L}\begin{bmatrix} 1 & -1 & 0 \\ -1 & (1+1/\alpha) & -1/\alpha \\ 0 & -1/\alpha & 1/\alpha \end{bmatrix}\begin{Bmatrix} c \\ c \\ c \end{Bmatrix}$$

Depending on the number of significant digits used in a numerical calculation, extreme values of α may lead to non-zero residual forces. In addition to the errors in the residual forces, we may want to compute a check on the potential energy

$$\mathbf{E} = \mathbf{R}^{\mathrm{T}}\mathbf{K}\mathbf{R} \stackrel{?}{=} \mathbf{0}$$

which should result in a 6×6 null matrix. Of course, these rigid body checks can be applied at the element level if desired. At the element level we usually also have access to the strain–displacement matrix, $\mathbf{B}$, so that

$$\boldsymbol{\varepsilon} = \mathbf{B}\mathbf{X}$$

where $\boldsymbol{\varepsilon}$ denotes the strain vector and $\mathbf{X}$ the element nodal displacements. Usually, we compute the element results

$$\mathbf{K} = \int \mathbf{B}^{\mathrm{T}}\mathbf{D}\mathbf{B}\,\mathrm{d}V$$

If we move the element as a rigid body, $\mathbf{X} = \mathbf{R}$, then the strain, $\boldsymbol{\varepsilon}$, should be zero at all points (including the integration points) in the element. Thus we might use

$$\boldsymbol{\varepsilon} = \mathbf{B}\mathbf{R} \stackrel{?}{=} \mathbf{0}$$

to try to assess a bad element before assembly begins. For the bar element used in the patch test, we know that $\mathbf{B} = [-1 \;\; 1]/L$ so that a rigid body x-translation, $\mathbf{R} = [c \;\; c]$, gives

$$\boldsymbol{\varepsilon} = \frac{1}{L}[-1 \;\; 1]\begin{Bmatrix} c \\ c \end{Bmatrix} = \mathbf{0}$$

as expected. For numerically integrated elements this could be checked at each quadrature point.

If we have problems involving mass distributions we may want to consider the kinetic energy. For a body translating with a constant speed it is simply one-half the mass times the speed squared, $T = mV^2/2$. Such a speed could be expressed as $\dot{\mathbf{x}} = \mathbf{R}/\Delta t$. This suggests that it could be useful to examine the product

$$\mathbf{T} = \frac{1}{2(\Delta t)^2}\mathbf{R}^{\mathrm{T}}\mathbf{M}\mathbf{R}$$

where **M** is the mass matrix. Some of the terms in **T** are translational energies, while others are energy due to the rotation about a fixed point as a rigid body. This could also be executed at the element level.

Another item of concern is the possibility of ill-conditioning of **K** and its condition number. If we return to the distorted patch assembly and restrain the third node, then **K** has the form

$$\mathbf{K} = \begin{bmatrix} K_A & -K_A \\ -K_A & (K_A + K_B) \end{bmatrix}$$

If the second element were very long, $\alpha \gg 1$, compared to the first then $K_A \gg K_B$, so that K_A dominates the structural stiffness, **K**. However, the inverse, $\mathbf{K}^{-1}$, is dominated by K_B, since

$$\mathbf{K}^{-1} = \frac{1}{K_B}\begin{bmatrix} \left(1 + \frac{K_B}{K_A}\right) & 1 \\ 1 & 1 \end{bmatrix}$$

This means that the displacements, and thus the stresses, are governed by K_B. This illustrates that a common cause of ill-conditioned problems in thin-walled structures is a large ratio of element stiffness to supporting structure stiffness. Similar problems often occur when large elements adjoin small ones. There is a trend to use 'pre-conditioned' iterative solvers to help overcome the accuracy problems of such formulations where the condition number of the system is large.

Each stiffness matrix **K**, will have a condition number, denoted here as $C(\mathbf{K})$. It has been shown that if p decimal digits per computer word are used to represent the stiffness terms, then the computed displacements will be correct to q decimal digits, where

$$q \geqslant p - \log_{10} C(\mathbf{K})$$

This log dependence on the condition number includes the effect of truncation errors. It does not include round-off errors. The calculation of the condition number can be expensive, so approximate values may be used. A typical range of values for finite element solutions is $10^3 < C(\mathbf{K}) < 10^6$. Its minimum value is unity. For a symmetric matrix, the condition number is defined as

$$C(\mathbf{K}) = \lambda_{max}/\lambda_{min}$$

where λ is an eigenvalue of a sealed version of **K**. The scaled version has diagonal coefficients that are unity. This is accomplished by using a diagonal matrix, **d**, defined from the diagonal of **K** as

$$\mathbf{d}_i = \mathbf{K}_{ii}^{-1/2}$$

Then the necessary scaled form is

$$\mathbf{K}_s = \mathbf{dKd}$$

The value of λ_{max} is often estimated to be two, since a number of tests have shown that $1{\cdot}5 < \lambda_{max} < 3{\cdot}2$. Thus one usually needs only to estimate the value of λ_{min}. This can be done by employing the Rayleigh method for 'natural frequencies'. An assumed mode shape with a single unknown amplitude is used to find λ_{min} from

$$|\mathbf{K} - \lambda_{min}\mathbf{M}| = \mathbf{0}$$

where the diagonal 'mass' matrix is defined as

$$\mathbf{M} = \mathbf{d}^{-1}\mathbf{d}^{-1}$$

As a final example, we will consider how to assess the behavior of a typical thin-walled structure element. Generally speaking, most analysts will utilize an existing computer program. These come with an element library that may offer several element types that can be applied to thin-walled structures. The most common programs are the public domain software, such as SAP, and the widely used commercial products, such as ANSYS, NASTRAN, PAFEC, etc. However, these useful libraries of elements may not give accurate and dependable results for thin-walled structures. When evaluating plates and shells, Robinson and Blockham[10] and Jones and Fong[15] found very serious errors in many elements. These types of studies showed that some elements should be avoided, and others only become inaccurate in certain load states. Two load states in particular were found to give very erroneous results for several elements. Both essentially result in a state of constant twist in the plate or shell elements. Here the single-element test will be utilized to demonstrate how one should

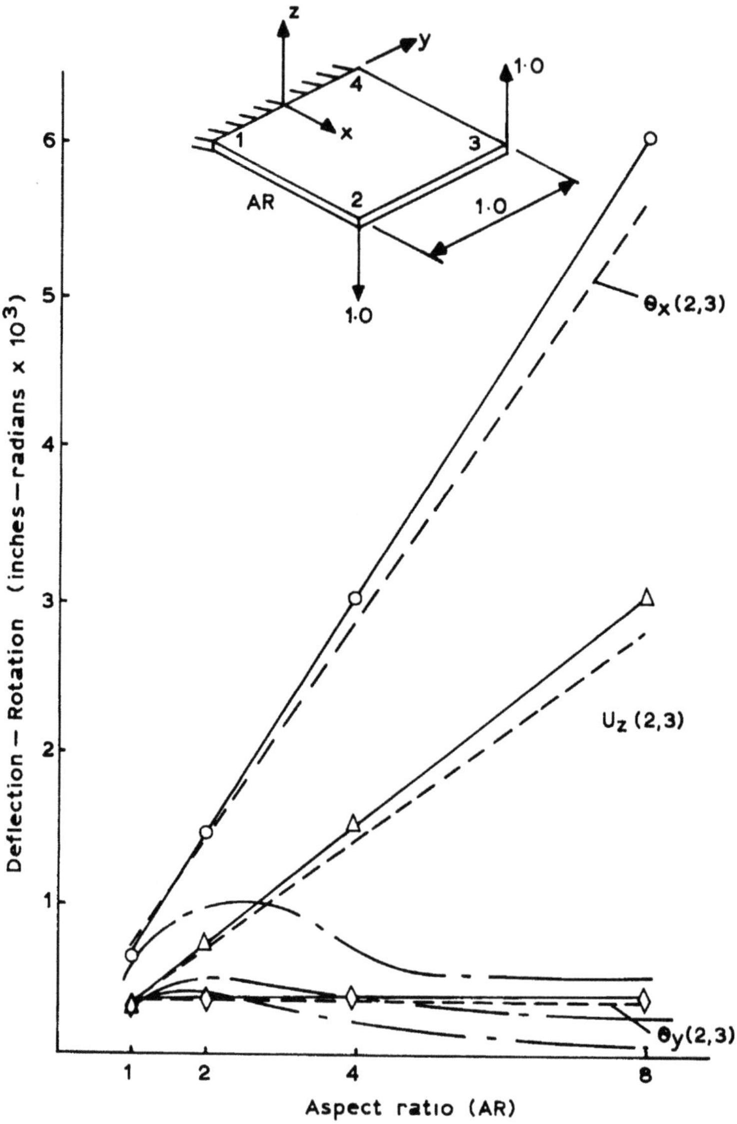

FIG. 3.4. Single element test for twisted elements. ———, Semi Loof. -----, Theory. —-—, QDPLT. Boundary conditions: $(\Theta_x)_1 = (\Theta_x)_4 = 0$, $(\Theta_y)_1 = (\Theta_y)_4 = 0$, $(U_z)_1 = (U_z)_4 = 0$.

attempt to evaluate elements to be used in a thin-walled structure. The loading condition will be a pair of equal and opposite forces causing a pure twist couple. The single element will have a square shape for the initial solution. That makes the aspect ratio unity. Most elements would give reasonable deflections and stresses in that state. Then, holding all other factors constant, the length of the element is increased to increase the aspect ratio. Many elements would then begin to give less accurate results. For this problem, a theoretical solution is available for any aspect ratio. In other cases one may need to employ a very fine mesh solution to use as a benchmark for the single element case. Figure 3.4 shows the loads, supports and the theoretical solutions (dashed) for the transverse deflection and the two rotations at a corner of the single element. Selecting a program and testing members of its element library, one can find results like those shown for the QDPLT element. Even for slightly distorted elements it gives 50% error in the deflections. At a reasonable aspect ratio of three it is in error by 100% or more. The error rapidly gets worse as the aspect ratio continues to increase.

Since a state of twist is common in thin-walled structures one should decide to select another member of the element library or to employ another program. Irons[4,16] has recommended the Semi Loof shell for use with thin-walled structures. It is a truly doubly curved surface element. However, it can also be used for flat plates such as the one in our current assessment. Repeating the single-element test with a Semi Loof element shows drastic improvement in both the transverse deflections and (extrapolated) rotations. This is true for the range of aspect ratios previously suggested as acceptable in our validation checks.

3.5. SUMMARY AND CONCLUSION

The use of finite elements to analyze thin-walled structures requires great care. One should not rely on programs as 'black boxes' to supply deflections, stresses etc. Each element type to be employed should be personally tested or researched so that one knows its limitations. Extensive validation checks should be utilized, even if one must write a special program to supplement an existing FEM code. Those wanting to research further the special difficulties associated with thin-walled structures should read the numerous works of the late Bruce Irons.

REFERENCES

1. AKIN, J. E., *Applications and Implementation of Finite Element Methods,* Academic Press, 1981.
2. COOK, R. D., *Concepts and Applications of Finite Element Analysis,* John Wiley, 1974.
3. HUGHES, O. F., *Ship Structural Design,* John Wiley, New York, 1983.
4. IRONS, B. M. and AHMAD, S., *Techniques of Finite Elements,* John Wiley, 1980.
5. AKIN, J. E., *Finite Element Analysis for Undergraduates,* Academic Press, 1986.
6. DUNDER, V. F. and BELONGOFF, G., Comparing finite element programs in engineering, *NASTRAN User Conference,* 1978.
7. NOOR, A. K. and PILKEY, W. D., *State of the Art Surveys on Finite Element Technology,* ASME Publ. H000290, New York, 1983, 530 pp.
8. IRONS, B. M. and RAZZAQUE, A., Experience with the patch test for convergence of the finite element method, in *Mathematical Foundation of the F.E.M.,* ed. A. R. Aziz, Academic Press, 1972, pp. 557–87.
9. ROBINSON, J., A single element test, *Comp. Math. Appl. Mech. Engng,* **7,** 1976, 191–200.
10. ROBINSON, J. and BLOCKHAM, S., *An Evaluation of Plate Bending Elements,* Robinson Ford Assoc., Dorset, England, 1981.
11. ROBINSON, J. and HAGGENMACHER, G. W., Element warning diagnostics, *Finite Element News,* Issues 3 & 4, 1982.
12. HENSHELL, R. D., *Programs for Automatic Finite Element Calculations—PAFEC, Their Theory and Results,* PAFEC Ltd, Strelley, Nottingham, England, 1976.
13. GJELSVIK, A., *Theory of Thin Walled Beams,* John Wiley, New York, 1981.
14. KAWAI, T., The application of finite element methods to ship structures, *Computers and Structures,* **3,** 1973, 1175–94.
15. JONES, J. W. and FONG, H. H., Evaluation of NASTRAN, in *Structural Mechanics Software Series IV,* eds N. Perrone *et al.,* University Press of Virginia, Charlottesville, VA, 1982, pp. 177–89.
16. IRONS, B. M., The Semi Loof shell element, in *Finite Elements for Thin Shells and Curved Members,* eds D. Ashwell and R. Gallagher, John Wiley, 1976, pp. 197–222.
17. ZIENKIEWICZ, O. C., *The Finite Element Method,* 3rd edn, McGraw-Hill, New York, 1979.
18. BARLOW, J., Optimal stress locations in finite element models, *Int. J. Num. Meth. Engng,* **10,** 1976, 241–51.
19. BAIER, H., Checking techniques for complex finite element models, in *Accuracy, Reliability and Training in FEM Technology,* ed. J. Robinson, Robinson & Associates, Wimborne, Dorset, U.K. 1984, pp. 145–56.

Chapter 4

The Analysis of Thin-walled Membrane Structures Using Finite Elements

C. T. F. Ross

Department of Mechanical Engineering,
Portsmouth Polytechnic, Portsmouth, UK

4.1. INTRODUCTION

Thin-walled membranes have been used for supporting loads for many centuries. Some of the earliest membrane structures were constructed in the jungles of Asia, where they appeared in the form of foot bridges, made from vines and from other undergrowth. Such structures resist lateral loads in tension, as vines, cables, etc., and have a negligible bending stiffness compared to their axial tensile stiffness (Fig. 4.1).

In the West, during the last century, the most spectacular use of membranes was made in the design and construction of suspension bridges, when cables were used to support the main body of the bridge, (Fig. 4.2). By placing the cables in a state of tension, the bridge is supported by vertical forces transmitted through the ties, so that bending effects on the bridges are minimised.

Two advantages of building a structure that resists its loads in tension are:

(a) A thin-walled structure is relatively stiff in tension, and if it is used in this manner it can have a high structural efficiency.
(b) The stiffness of the membrane usually increases with increasing tension, until its material of construction yields.

Typical examples of structures whose stiffness increases with tension are footballs, basketballs, etc., where the introduction of internal pneumatic pressure causes the balls to become relatively hard.

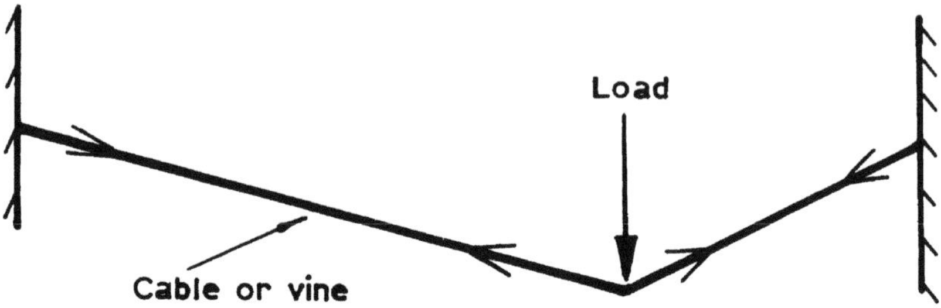

FIG. 4.1. Cable or vine under lateral load.

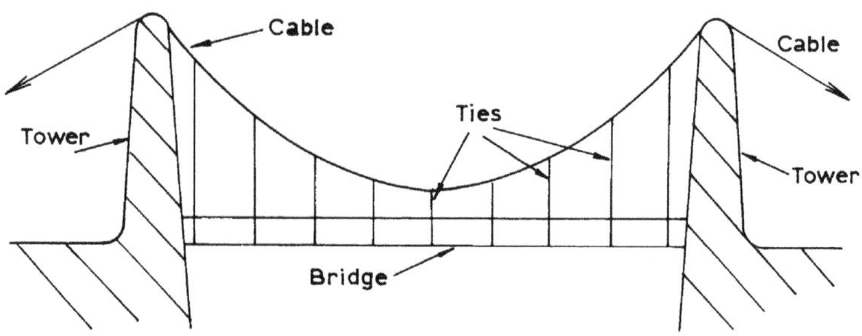

FIG. 4.2. A suspension bridge.

4.1.1. Today, membranes appear in many forms, from wire cables to dams and from tyres to sports' stadiums. A brief description of these structures will now be given.

4.2. THE USE OF THIN-WALLED MEMBRANE STRUCTURES

Major membrane structures come under two categories;

(a) air-supported structures and
(b) water-supported structures.

4.3. AIR-SUPPORTED STRUCTURES

The first major type of air-supported structure appeared as an airship. The most famous was built and tested by Count Zeppelin in 1900.

4.3.1. Airships can be described as power-driven streamlined balloons, and were inflated with hydrogen during their earlier development. In 1930, the British airship, the R101, was destroyed by fire, with serious loss of life and, as a result, the use of airships became temporarily 'out of favour'.

In fact, it was not until after the late 1940's, when large quantities of helium became available, that interest in these airship structures was revived. Advantages of airships are that they can lift enormous loads, make little noise during flight, and have a low fuel consumption. In addition, they do not require special airstrips or sites.

4.3.2. Another air-supported vehicle is the hovercraft, which was invented by Sir Christopher Cockerell. These vehicles ride on a cushion of air, which is partially enclosed by a membrane skirt attached to its peripheral base. The first hovercraft service was of an experimental nature and operated between Rhyl and Wallasey in the UK, during the summer of 1962.

4.3.3. According to Newman *et al.*,[1] the feasibility of using an air-inflated structure as a building was first conceived by Lanchester in 1917,[2] when he patented plans for a field hospital.

In 1950, the first major air-inflated structure to be used as a building was designed and manufactured, when Bird[3] produced his spherical radomes to house early-warning radar antennae.

Since then, many air-inflated structures have appeared all over the world, and Newman *et al.*,[1] quote the following buildings, which are among the most famous:

(a) The Pontiac Silverdome stadium in Detroit, which has a height of 60 m and a base area of 40 000 m^2.
(b) The stadium at British Columbia Place, which has a height of 30 m and an elliptical base of dimensions 220 × 168 m.
(c) The stadium at Dalhousie University,[4] which has a height of 11 m and an elliptical base of dimensions 86 × 66 m.
(d) The Pan American spherical membrane dome at the Brussels World Fair in 1958.
(e) The spherical dome used by the U.S. Atomic Energy Commission in their travelling exhibition, namely, 'Atoms for Peace', which had a dome height to base diameter ratio of 0·4.

Typical values for dome height to base diameter ratios for spherical domes vary from 0·33 to 0·5 and, very often, the base diameter of these domes exceeds 100 m.

The material of construction of such buildings is usually thin and impervious, with little flexural stiffness. On some occasions, particularly for larger structures, the impervious material may be reinforced with cables.

In the case of the Dalhousie University stadium,[4] the requirement was to construct a building with an elliptical base of dimensions 86×66 m and of height 11 m and, to achieve this, it was necessary to use a type 304 stainless steel membrane of thickness 1·59 mm.

The normal internal design operating pressure was 372·3 Pa with the deadweight of the membrane being 124·5 Pa, so that after deducting the membrane deadweight and the insulating ceiling weight, etc., the net tensioning pressure was approximately 191·5 Pa. This pressure was insufficient, particularly to resist snow loads of about 1293 Pa, and a compression ring was fitted which caused the net uplift to be increased to about 479 Pa.

An aero-elastic pressurised model of the Dalhousie roof of scale 1:330 was tested in a wind tunnel, under turbulent boundary layer flow conditions, which found the membrane roof to be aerodynamically stable up to wind speeds of 209 km h^{-1}.

According to Springfield and Sinorski,[4] the maximum tensile stress, as predicted by a NASTRAN program, did not exceed 186 MPa. The 0·2%, strength and the ultimate tensile strength of the material of construction were 262 MPa and 579 MPa respectively. Further details of this building, including its fabrication and erection, are given by Springfield and Sinorski.

For tall membrane structures, if the internal uplift is insufficient, buckling of the structure can be caused by high winds,[1] but the more usual method of failure is due to local tearing of the membrane.

4.3.4. Another application of a temporary air-supported membrane was used by Northamptonshire County Council during the construction of a reinforced concrete dome of 30 m diameter and height 11 m. In this case, the engineers built the web of the required steel reinforcement, and inserted a temporary inflated membrane within this structure (Fig. 4.3). The reinforced concrete dome was then constructed by spraying liquid concrete over the membrane and steel reinforcement. Later, the membrane dome was deflated and removed.

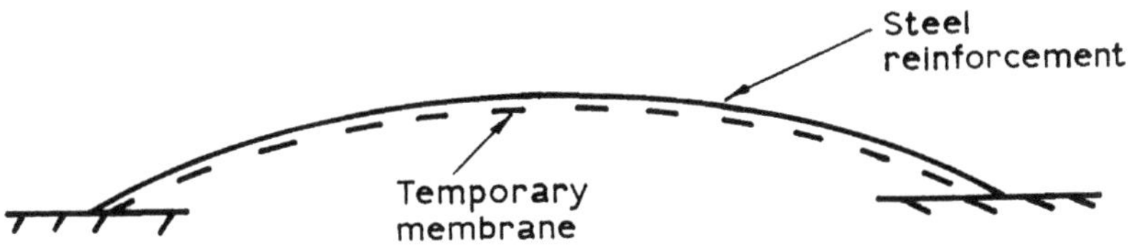

FIG. 4.3. Membrane used by Northamptonshire County Council.

4.4. WATER-INFLATED MEMBRANES

One of the most useful water-inflated membranes, which can also be partially air-inflated, is the membrane dam, invented by N. M. Imbertson and L. E. Tabor of Los Angeles.[5]

Water-inflated membrane dams are inexpensive methods of controlling floods and can also be used as a means for temporary storage of water (Fig. 4.4). The membrane dams of Fig. 4.4 can be deflated or inflated, as required.

4.4.1. The membrane dam of Imbertson and Tabor was called a 'Fabridam', and was used to increase the height of the Koombooloomba dam on the Upper Tully river in 1965.

The spillway of the Koombooloomba dam had a length 60·96 m while the maximum design height of the 'Fabridam' was 1·52 m. The dam could be inflated or deflated, through the use of a pump, where water was used as the pressure raising or lowering medium. The

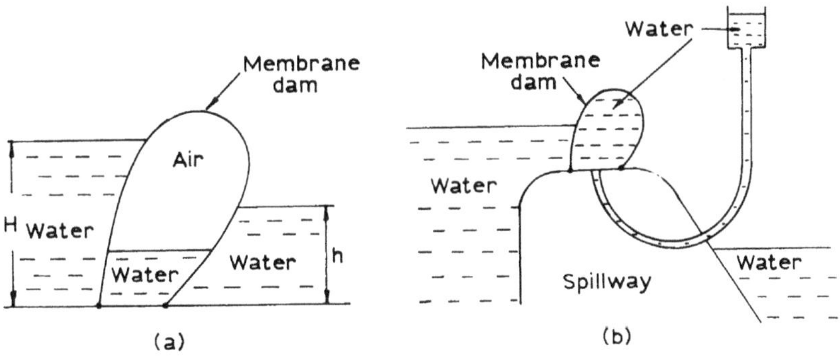

FIG. 4.4. Water- and air-inflated membrane dams. (a) Water- and air-inflated dam; (b) water-inflated dam on a spillway crest.

cross-section of the 'Fabridam' is shown diagrammatically in Fig. 4.4(b), where it can be seen that it was firmly secured to the top of the spillway of the Koombooloomba dam.

Although the maximum height of the 'Fabridam' was only 1·52 m, it provided sufficient additional water to the Kareeya Hydro-Electric Power Station that the cost of manufacture and installation of the dam was recovered within two years of its commencement of operation.

The internal head that the 'Fabridam' was allowed to sustain was 2·26 m and, as the maximum design height was 1·52 m, the additional head of 0·74 m was used to put the envelope into a state of hoop tension.

The material of the 'Fabridam' had a weight of 4·5 kg m^{-2} and was similar to that used for car tyres. It was constructed by the Firestone Tire and Rubber Company, in the form of a long thin tube.

Details of the installation, testing and operation of the 'Fabridam' are given by Shepherd *et al.*[5]

4.4.2. Since then, a number of membrane dams have been used for controlling water flow, and Harrison[6] gives an example of a water-inflated membrane dam where:

$H = 5{\cdot}79$ m, $h = 1{\cdot}52$ m, base width $= 6{\cdot}1$ m,

curved perimeter $= 18{\cdot}3$ m, wall thickness $= 7{\cdot}62$ mm, and

elastic modulus $= 2{\cdot}07 \times 10^9$ N m^{-2}.

4.5. FINITE ELEMENT ANALYSIS OF MEMBRANE DAMS

Unless the membrane is in the form of a thin curved shell, where deflections are small, the effects of large displacements must be considered. The load–displacement relationship will therefore be of the form:

$$\{q\} = ([K] + [K_G])\{u\} \tag{4.1}$$

where

$\{q\}$ = a vector of external loads acting on the entire structure $\Sigma\,\{P^0\}$

$[K]$ = 'small deflection' system stiffness matrix in global co-ordinates $\Sigma\,[k^0]$

$[K_G]$ = geometrical stiffness matrix of the system in global co-ordinates $\Sigma\, [k_G^0]$
$[k^0]$ = 'small deflection' elemental stiffness matrix in global coordinates $[\Xi]^T[k][\Xi]$
$[k_G^0]$ = geometrical stiffness matrix of the element in global co-ordinates
$\{P^0\}$ = a vector of elemental nodal forces in global co-ordinates
$[k]$ = 'small deflection' elemental stiffness matrix in local co-ordinates
$[k_G]$ = geometrical stiffness matrix of the element in local co-ordinates
$\{P\}$ = a vector of elemental nodal forces in local co-ordinates
$[\Xi]$ = a matrix of directional cosines

More precise definitions of these equations and matrices are given in refs 7–9.

4.5.1. The load–displacement relationship for a membrane structure is likely to be of the form shown in Fig. 4.5, where it can be seen that the stiffness of the membrane increases with displacement, until the material yields.

When the membrane is just inflated, and in a state of zero tension,

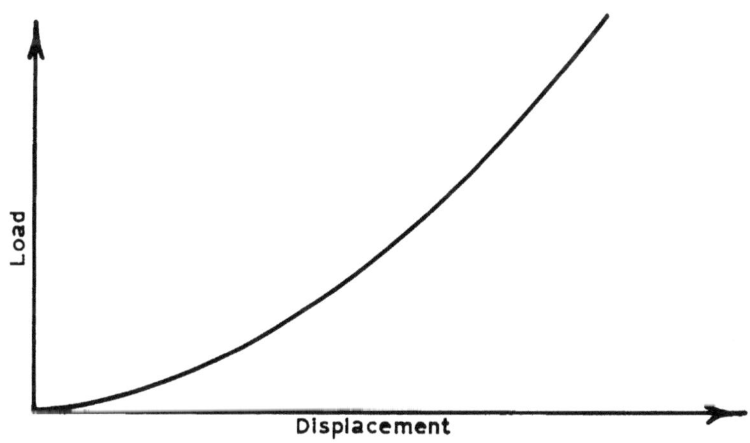

FIG. 4.5. Load–displacement relationship for a membrane under tension.

$[K_G]$ will be null, so that, if a 'small load' $\{\delta q_1\}$ is applied, then

$$\{\delta u_1\} = [K^{-1}]\{\delta q_1\} \tag{4.2}$$

where $\{\delta u_1\}$ is a vector of displacements due to $\{\delta q_1\}$. Knowing $\{\delta u_1\}$, the tensile stresses in the membrane can be calculated and hence $[K_G]$, which is dependent on initial stresses, can be determined. The next stage is to determine the change in geometry of the membrane, if any, so that the new 'small deflection' stiffness matrix, $[K_1]$, can be determined, and also to apply another 'small load' $\{\delta q_2\}$, so that

$$\{\delta u_2\} = ([K_1] + [K_G])^{-1}\{\delta q_2\} \tag{4.3}$$

Both $\{\delta u\}$ and $\{\delta q\}$ from eqns (4.2) and (4.3) can now be added together, as indeed can the stresses from these two equations, to obtain the new $[K_2]$ and $[K_{G1}]$. Similarly, by adding another incremental load, $\{\delta q_3\}$, the incremental displacements, $\{\delta u_3\}$, can be determined from:

$$\{\delta u_3\} = ([K_2] + [K_{G1}])^{-1}\{\delta q_3\} \tag{4.4}$$

Eventually, for the nth stage,

$$\{\delta u_n\} = ([K_{(n-1)}] + [K_{G(n-2)}])\{\delta q_n\} \tag{4.5}$$

where

$$\{q\} = \sum_{i=1}^{n} \{\delta q_i\} \tag{4.6}$$

$$\{u\} = \sum_{i=1}^{n} \{\delta u_i\} \tag{4.7}$$

The process is a long and involved one, and is very demanding on computer time. Many of today's designs of membrane structures are approximate, purely to save computational time. However, in the future, such analyses will be more involved as computational power continues to grow, and becomes much less expensive. A number of finite elements will now be described, which can be used for one-, two- and three-dimensional analyses.

4.6. ONE-DIMENSIONAL ELEMENT

One-dimensional elements can be used for analysing cables, inflatable membrane dams and also, flat two-dimensional membranes.

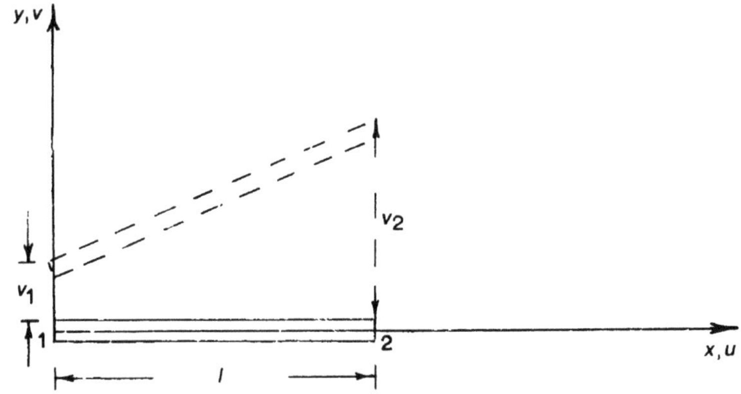

FIG. 4.6. One-dimensional rod.

4.6.1. The simplest one-dimensional element is the rod shown in Fig. 4.6, the geometrical stiffness matrix of which is given by:

$$[k_G] = \frac{F}{l}\begin{bmatrix} 1 & -1 \\ -1 & 1 \end{bmatrix} \begin{matrix} v_1 \\ v_2 \end{matrix} \quad \text{(columns } v_1, v_2\text{)} \tag{4.8}$$

where F is the internal force in the rod (tensile +ve). Consider the cable of Fig. 4.7, whose weight to length ratio is unity, so that the vector of elemental forces is

$$\{P^0\} = \begin{Bmatrix} -l/2 \\ -l/2 \end{Bmatrix}$$

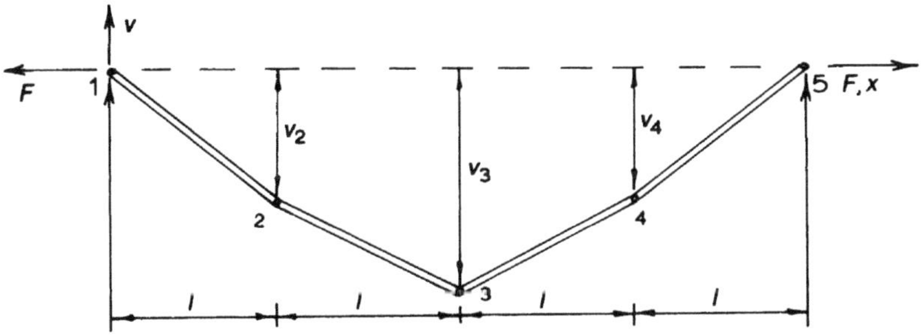

FIG. 4.7. Hanging cable.

Assuming that the deflections are small, such that $u_1 = u_2 = u_3 = u_4 = u_5 = 0$, it can be shown that the force–displacement relationship for the entire structure, is given by:

$$\begin{Bmatrix} -l \\ -l \\ -l \end{Bmatrix} = \frac{F}{l} \begin{bmatrix} 2 & -1 & 0 \\ -1 & 2 & -1 \\ 0 & -1 & 2 \end{bmatrix} \begin{Bmatrix} v_2 \\ v_3 \\ v_4 \end{Bmatrix} \quad (4.9)$$

But,

$$v_2 = v_4$$

Therefore, eqn (4.9) reduces to the simpler form of:

$$\begin{Bmatrix} -l \\ -l \end{Bmatrix} = \frac{F}{l} \begin{bmatrix} 2 & -1 \\ -2 & 2 \end{bmatrix} \begin{Bmatrix} v_2 \\ v_3 \end{Bmatrix} \quad (4.10)$$

Therefore:

$$\begin{Bmatrix} v_2 \\ v_3 \end{Bmatrix} = \frac{-l^2}{F} \begin{bmatrix} 2 & -1 \\ -2 & 2 \end{bmatrix} \begin{Bmatrix} 1 \\ 1 \end{Bmatrix}$$

$$= \frac{-l^2}{F} \begin{Bmatrix} 1{\cdot}5 \\ 2 \end{Bmatrix} \quad (4.11)$$

According to Sack,[10] the second-order differential equation governing the shape of a free-hanging cable is given by:

$$\mathrm{d}^2 v/\mathrm{d}x^2 = -1/F \quad (4.12)$$

or:

$$v = -\frac{x^2}{2F} + C_1 x + C_2$$

At $x = 0$, $v = 0$ and therefore $C_2 = 0$. At $x = 4l$, $v = 0$ and therefore $C_1 = 2l/F$; i.e.

$$v = -\frac{x^2}{2F} + \frac{2lx}{F}$$

giving

and

$$\left.\begin{aligned} v_2 &= 1{\cdot}5l^2/F \\ v_3 &= 2l^2/F \end{aligned}\right\} \text{ as required}$$

These nodal values are exactly the same as the finite element solution, except that the latter underestimates v at points other than the nodes, as shown in Fig. 4.8.

4.6.2. For large deflections, a better one-dimensional element than that of section 4.6.1, is the beam–column element of Fig. 4.9, in which

$$[k]=\frac{AE}{l}\begin{array}{c} \begin{array}{cccccc} u_1 & v_1 & \theta_1 & u_2 & v_2 & \theta_2 \end{array} \\ \left[\begin{array}{cccccc}
c^2+\frac{s^2\cdot 12z}{l^2} & & & & & \\
cs-\frac{cs\cdot 12z}{l^2} & s^2+c^2\cdot\frac{12z}{l^2} & & \text{Symmetrical} & & \\
\frac{s\cdot 6z}{l} & -c\cdot\frac{6z}{l} & 4z & & & \\
-c^2-s^2\cdot\frac{12z}{l^2} & -cs+cs\cdot\frac{12z}{l^2} & -\frac{s\cdot 6z}{l} & c^2+\frac{s^2\cdot 12z}{l^2} & & \\
-cs+\frac{cs\cdot 12z}{l^2} & -s^2-\frac{c^2\cdot 12z}{l^2} & \frac{c\cdot 6z}{l} & cs-\frac{cs\cdot 12z}{l^2} & s^2+\frac{c^2\cdot 12z}{l^2} & \\
s\cdot\frac{6z}{l} & -c\cdot\frac{6z}{l} & 2z & -s\cdot\frac{6z}{l} & \frac{c\cdot 6z}{l} & 4z
\end{array}\right]\end{array}\begin{array}{c} u_1 \\ v_1 \\ \theta_1 \\ u_2 \\ v_2 \\ \theta_2 \end{array}$$

$$[k_G]=\frac{F}{l}\begin{array}{c} \begin{array}{cccccc} u_1^0 & v_1^0 & \theta_1 & u_2^0 & v_2^0 & \theta_2 \end{array} \\ \left[\begin{array}{cccccc}
\frac{6s^2}{5} & & & & & \\
-\frac{6cs}{5} & \frac{6c^2}{5} & & & & \\
\frac{ls}{10} & -\frac{lc}{10} & \frac{2l^2}{15} & & & \\
-\frac{6s^2}{5} & \frac{6cs}{5} & -\frac{ls}{10} & \frac{6s^2}{5} & & \\
\frac{6cs}{5} & \frac{6c^2}{5} & \frac{lc}{10} & -\frac{6cs}{5} & \frac{6c^2}{5} & \\
\frac{ls}{10} & -\frac{lc}{10} & -\frac{l^2}{30} & -\frac{ls}{10} & \frac{lc}{10} & \frac{2l^2}{15}
\end{array}\right]\end{array}\begin{array}{c} u_1^0 \\ v_1^0 \\ \theta_1 \\ u_2^0 \\ v_2^0 \\ \theta_2 \end{array}$$

where $z = EI/(AE)$, F is the axial force in the beam (+ve if tensile), E is the elastic modulus, A is the cross-sectional area, I is the 2nd

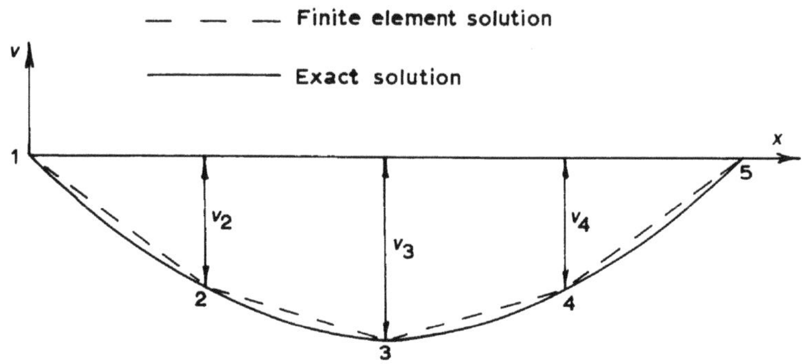

FIG. 4.8. Comparison between finite element and 'exact' solutions.

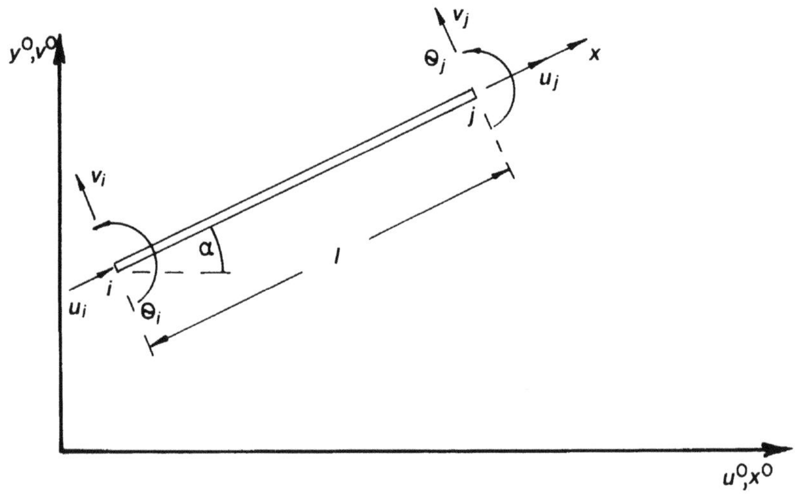

FIG. 4.9. Beam-column element.

moment of area (this must be very small, say, $A/1E20$), l is the elemental length, $c = \cos \alpha$, and $s = \sin \alpha$.

4.7. TWO-DIMENSIONAL ELEMENTS

Suitable two-dimensional elements may include any one of the following:

(a) the three-node simplex triangular element of Turner *et al.*;[11]

(b) the four-node quadrilateral of Taig;[12]
(c) the eight-node isoparametric element of Ahmad *et al.*[13]

The main advantage of the three-node triangular element is that it is easy to generate, but because it assumes constant strains over its surface, it is not very accurate for practical applications.

In this context, the four-node quadrilateral is superior to the three-node triangle, as it assumes linear variations in strain across its surface, but it requires numerical integration.

The eight-node quadrilateral assumes a second-order variation in strain, but it too requires numerical integration. However, if reduced integration is used, it can be economical on computational time and costs.

Prior to obtaining the three-node triangular and four-node quadrilateral 'shell' elements, it will be necessary to obtain their in-plane parent elements.

4.8. IN-PLANE THREE-NODE TRIANGULAR ELEMENT

This element, which is the constant strain triangular element of Turner *et al.*,[11] is shown in Fig. 4.10. The element is described by three corner nodes, with two degrees of freedom per node, namely, u^0 and v^0.

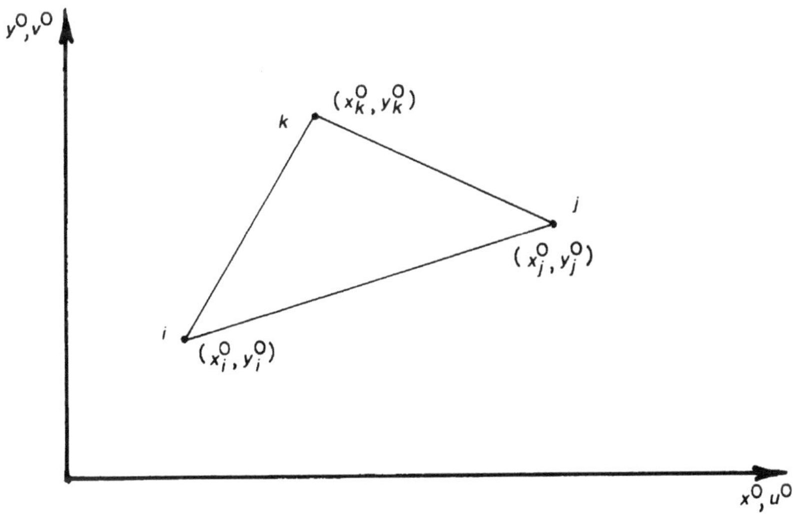

FIG. 4.10. In-plane triangular element.

The small-deflection stiffness matrix $[k]$ for the element of Fig. 4.10 is given by:

$$[k]=\frac{E't}{4\Delta}\left[\begin{array}{ccc|ccc}
u_i^0 & u_j^0 & u_k^0 & v_i^0 & v_j^0 & v_k^0 \\
b_ib_i+\gamma C_iC_i & & & \text{symmetrical} & & \\
b_ib_j+\gamma C_iC_j & b_j^2+\gamma C_j^2 & & & & \\
b_ib_k+\gamma C_iC_k & b_jb_k+\gamma C_jC_k & b_k^2+\gamma C_k^2 & & & \\
\hline
\mu b_iC_i+\gamma b_iC_i & \mu b_jC_i+\gamma b_iC_j & \mu b_kC_i+\gamma b_iC_k & C_i^2+\gamma b_ib_i & & \\
\mu b_iC_j+\gamma b_jC_i & \mu b_jC_j+\gamma b_jC_j & \mu b_kC_j+\gamma b_jC_k & C_iC_j+\gamma b_ib_j & C_j^2+\gamma b_j^2 & \\
\mu b_iC_k+\gamma b_kC_i & \mu b_jC_k+\gamma b_kC_j & \mu b_kC_k+\gamma b_kC_k & C_iC_k+\gamma b_ib_k & C_jC_k+\gamma b_jb_k & C_k^2+\gamma b_k^2
\end{array}\right] \tag{4.13}$$

$$a_i = x_j^0y_k^0 - x_k^0y_j^0$$
$$b_i = y_j^0 - y_k^0$$
$$C_i = x_k^0 - x_j^0$$
$$a_j = x_k^0y_i^0 - y_k^0x_i^0$$
$$b_j = y_k^0 - y_i^0$$
$$C_j = x_i^0 - x_k^0$$
$$a_k = x_i^0y_j^0 - x_j^0y_i^0$$
$$b_k = y_i^0 - y_j^0$$
$$C_k = x_j^0 - x_i^0$$
$$E' = E/(1-\nu^2)$$
$$\mu = \nu$$
$$\gamma = (1-\nu)/2$$

E = elastic modulus

ν = Poisson's ratio

t = plate thickness

$$\Delta = \tfrac{1}{2}\begin{vmatrix} 1 & x_i^0 & y_i^0 \\ 1 & x_j^0 & y_j^0 \\ 1 & x_k^0 & y_k^0 \end{vmatrix} = \text{area of triangle}$$

4.8.1. To obtain the geometrical stiffness matrix, the vector of additional strains, $\{\varepsilon_L\}$, due to large deflections, must be considered. This vector can be readily shown to be:

$$\{\varepsilon_L\} = \tfrac{1}{2}\begin{Bmatrix} \left(\dfrac{\partial w}{\partial x^0}\right)^2 \\ \left(\dfrac{\partial w}{\partial y^0}\right)^2 \\ 2\dfrac{\partial w}{\partial x^0}\dfrac{\partial w}{\partial y^0} \end{Bmatrix}$$

or

$$\begin{Bmatrix} \partial w/\partial x^0 \\ \partial w/\partial y^0 \end{Bmatrix} = [G]\begin{Bmatrix} w_i \\ w_j \\ w_k \end{Bmatrix}$$

where w_i, w_j and w_k are nodal deflections perpendicular to the plane of the plate, w is the deflection perpendicular to the plane of the plate at (x^0, y^0)

$$= \lfloor N_i \;\; N_j \;\; N_k \rfloor \begin{Bmatrix} w_i \\ w_j \\ w_k \end{Bmatrix}$$

$$N_i = (a_i + b_i x^0 + C_i y^0)/2\Delta$$

$$N_j = (a_j + b_j x^0 + C_j y^0)/2\Delta$$

and

$$N_k = (a_k + b_k x^0 + C_k y^0)/2\Delta$$

The geometrical stiffness matrix, $[k_G]$ is given by:

$$[k_G] = \int [G]^{\mathrm{T}}[\sigma][G]\,\mathrm{d}x^0\,\mathrm{d}y^0$$

$$[\sigma] = \begin{bmatrix} \sigma_x & \tau_{xy} \\ \tau_{xy} & \sigma_y \end{bmatrix}$$

where σ_x is the stress in the x^0 direction, σ_y is the stress in the y^0

direction, and τ_{xy} is the shear stress in the x^0–y^0 plane. Hence,

$$[k_G] = \frac{t}{4\Delta} \begin{bmatrix} \overset{w_i^0}{(b_i(\sigma_x b_i + \tau_{xy} C_i) + C_i(\tau_{xy} b_i + \sigma_y C_i))} & \overset{w_j^0}{} & \overset{w_k^0}{\text{symmetrical}} \\ (b_j(\sigma_x b_i + \tau_{xy} C_i) + C_j(\tau_{xy} b_i + \sigma_y C_i)) & (b_j(\sigma_x b_j + \tau_{xy} C_j) + C_j(\tau_{xy} b_j + \sigma_y C_j)) & \\ (b_k(\sigma_x b_i + \tau_{xy} C_i) + C_k(\tau_{xy} b_i + \sigma_y C_i)) & (b_k(\sigma_x b_j + \tau_{xy} C_j) + C_k(\tau_{xy} b_j + \sigma_y C_j)) & (b_k(\sigma_x b_k + \tau_{xy} C_k) + C_k(\tau_{xy} b_k + \sigma_y C_k)) \end{bmatrix} \quad (4.14)$$

4.9. 'SHELL' ELEMENTS

To analyse the curved membrane, it will be necessary to assemble eqn (4.13) into a 'shell' element of order 9×9. First, however, it is convenient to display the in-plane plate element of Fig. 4.10 in the local co-ordinate system of Fig. 4.11.

From Fig. 4.11, it follows

$$\begin{array}{lll} a_i = x_j y_k, & a_j = 0, & a_k = 0 \\ b_i = -y_k, & b_j = y_k, & b_k = 0 \\ C_i = x_k - x_j, & C_j = -x_k, & C_k = x_j \end{array}$$

The elemental stiffness matrix of the 'flat shell' element, $[k_f]$ is given by:

$$[k_f] = [k]_{9\times 9} \quad (4.15)$$

where $[k]_{9\times 9}$ has been expanded to include three columns and rows of zeros, corresponding to w_i, w_j and w_k.

It is now necessary to display the 'flat shell' element in three dimensional global co-ordinates, as shown in Fig. 4.12 but, to avoid

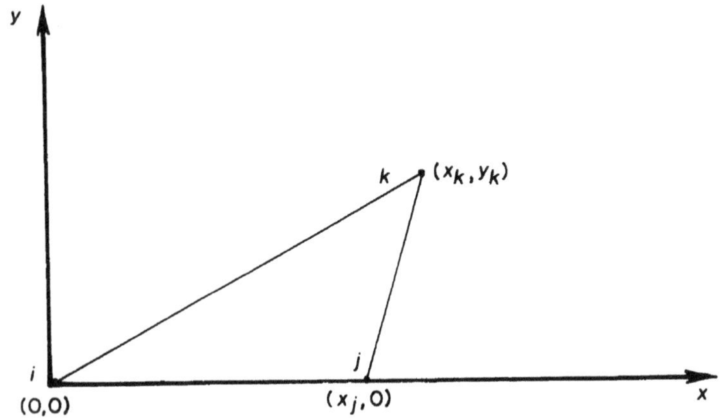

FIG. 4.11. In-plane triangular element in local co-ordinates.

numerical instability, small stiffnesses should be introduced in the direction of the w displacements. These stiffnesses can be made equal to very small values of the diagonal stiffnesses in the direction of the u and v displacements, rather similar to the method used by Zienkiewicz[7] for doubly curved shells.

The stiffness matrix for the 'shell' element, $[k_s]_{9\times 9}$, is given by:

$$[k_s] = [\Xi]^T [k]_{9\times 9} [\Xi] \tag{4.16}$$

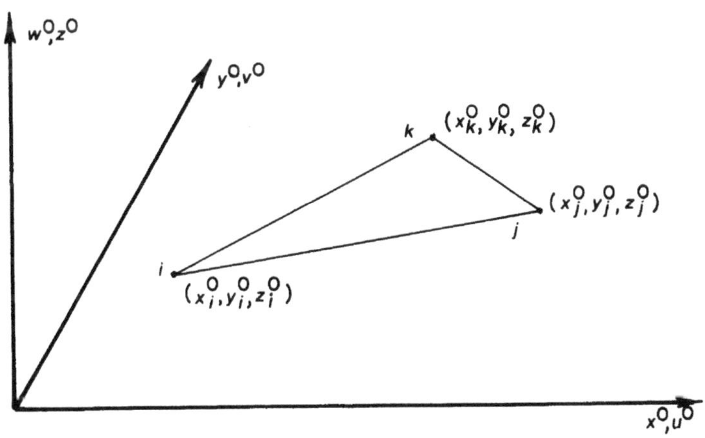

FIG. 4.12. 'Flat shell' element in global co-ordinates.

where

$$[\Xi] = \left[\begin{array}{c|c|c} \zeta & 0_3 & 0_3 \\ \hline 0_3 & \zeta & 0_3 \\ \hline 0_3 & 0_3 & \zeta \end{array}\right] \tag{4.17}$$

$$[\zeta] = \begin{bmatrix} C_{x,x^0} & C_{x,y^0} & C_{x,z^0} \\ C_{y,x^0} & C_{y,y^0} & C_{y,z^0} \\ C_{z,x^0} & C_{z,y^0} & C_{z,z^0} \end{bmatrix} \tag{4.18}$$

where 0_3 is a null matrix of order 3×3, C_{x,x^0} is the direction cosine of x with x^0, C_{x,y^0} is the direction cosine of x with y^0, and C_{x,z^0}, etc. are the direction cosines of x with z^0 etc.

Petyt[14] has shown that

$$\lfloor C_{x,x^0} \ C_{x,y^0} \ C_{x,z^0} \rfloor^{\mathrm{T}} = (\{\psi_j\} - \{\psi_i\})/l_{ji}$$

$$\lfloor C_{y,x^0} \ C_{y,y^0} \ C_{y,z^0} \rfloor = \lfloor \psi_{34} \rfloor / l_{34}$$

$$\lfloor C_{z,x^0} \ C_{z,y^0} \ C_{z,z^0} \rfloor = \frac{1}{\Delta'} \lfloor \Delta_{y^0,z^0} \ \Delta_{z^0,x^0} \ \Delta_{x^0,y^0} \rfloor$$

where

$$\{\psi_i\}^{\mathrm{T}} = \lfloor x_i^0 \ y_i^0 \ z_i^0 \rfloor$$

$$\{\psi_j\}^{\mathrm{T}} = \lfloor x_j^0 \ y_j^0 \ z_j^0 \rfloor$$

$$l_{ji}^2 = (\lfloor \psi_{ji} \rfloor \lfloor \psi_{ji} \rfloor^{\mathrm{T}})$$

$$l_{34}^2 = \lfloor \psi_{34} \rfloor \{\psi_{34}\}$$

$$\{\psi_{34}\} = [\lceil I \rfloor - \{\zeta_1\} \lfloor \zeta_1 \rfloor]\{\psi_{ki}\}$$

$$\lfloor \zeta_1 \rfloor = \frac{1}{l_{21}} \{\psi_{ji}\}^{\mathrm{T}}$$

$$l_{21} = \{\psi_{ji}\}^{\mathrm{T}} \{\psi_{ji}\}$$

$$\{\psi_{ki}\} = \begin{Bmatrix} x_k^0 - x_i^0 \\ y_k^0 - y_i^0 \\ z_k^0 - z_i^0 \end{Bmatrix}$$

$$\Delta' = \text{area of triangle } ijk$$

$$= [\Delta_{y^0,z^0}^2 + \Delta_{z^0,x^0}^2 + \Delta_{x^0,y^0}^2]^{1/2}$$

$$\Delta_{x^0,y^0} = \text{projected area of } \Delta_{ijk} \text{ on the } x^0\text{–}y^0 \text{ plane}$$

$$= \frac{1}{2}\begin{vmatrix} x_i^0 & y_i^0 & 1 \\ x_j^0 & y_j^0 & 1 \\ x_k^0 & y_k^0 & 1 \end{vmatrix}$$

$$\Delta_{y^0,z^0} = \text{projected area of } \Delta_{ijk} \text{ on the } y^0\text{–}z^0 \text{ plane}$$

$$= \frac{1}{2}\begin{vmatrix} y_i^0 & z_i^0 & 1 \\ y_j^0 & z_j^0 & 1 \\ y_k^0 & z_k^0 & 1 \end{vmatrix}$$

and

$$\Delta_{z^0,x^0} = \text{projected area of } \Delta_{ijk} \text{ on the } x^0\text{–}z^0 \text{ plane}$$

$$= \frac{1}{2}\begin{vmatrix} z_i^0 & x_i^0 & 1 \\ z_j^0 & x_j^0 & 1 \\ z_k^0 & x_k^0 & 1 \end{vmatrix}$$

4.10. FOUR-NODE QUADRILATERAL ELEMENT

This element, due to Taig,[12] allows a linear variation in strain, and is a more sophisticated element than the three-node constant strain triangular element of Turner *et al.*[11]

The element is shown in Fig. 4.13. The assumed displacement

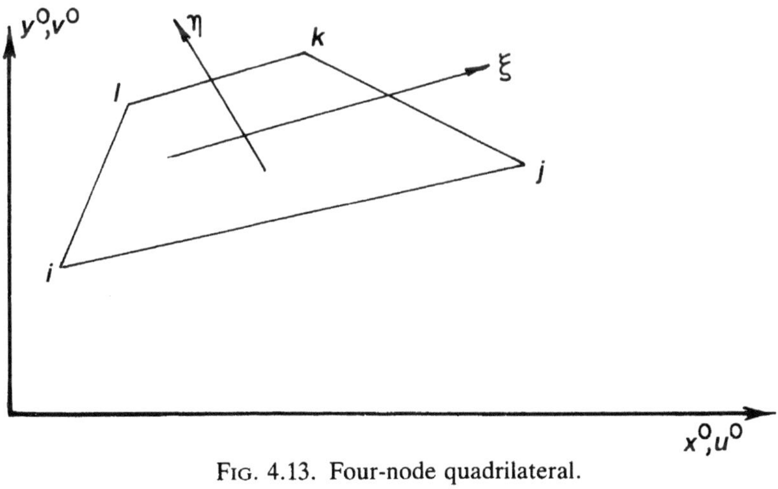

FIG. 4.13. Four-node quadrilateral.

functions are:

$$u^0 = \alpha_1 + \alpha_2 x^0 + \alpha_3 y^0 + \alpha_4 x^0 y^0 \tag{4.19}$$

$$v^0 = \alpha_5 + \alpha_6 x^0 + \alpha_7 y^0 + \alpha_8 x^0 y^0 \tag{4.20}$$

Ergatoudis *et al.*[15] have shown that the displacement assumptions of eqns (4.21) and (4.22) can be put in the form of the matrix equation:

$$\begin{Bmatrix} u^0 \\ v^0 \end{Bmatrix} = \begin{bmatrix} N_1 & 0 & N_2 & 0 & N_3 & 0 & N_4 & 0 \\ 0 & N_1 & 0 & N_2 & 0 & N_3 & 0 & N_4 \end{bmatrix} \begin{Bmatrix} u_1^0 \\ v_1^0 \\ u_2^0 \\ \downarrow \\ u_4^0 \\ v_4^0 \end{Bmatrix} \tag{4.21}$$

$$= [N]\{u_i^0\}$$

Similarly,

$$\begin{Bmatrix} x^0 \\ y^0 \end{Bmatrix} = \begin{bmatrix} N_1 & 0 & N_2 & 0 & N_3 & 0 & N_4 & 0 \\ 0 & N_1 & 0 & N_2 & 0 & N_3 & 0 & N_4 \end{bmatrix} \begin{Bmatrix} x_1^0 \\ y_1^0 \\ x_2^0 \\ \downarrow \\ y_4^0 \end{Bmatrix} \tag{4.22}$$

where $[N]$ is a matrix of shape functions, with

$$N_1 = \tfrac{1}{4}(1-\xi)(1-\eta)$$
$$N_2 = \tfrac{1}{4}(1+\xi)(1-\eta)$$
$$N_3 = \tfrac{1}{4}(1+\xi)(1+\eta)$$
$$N_4 = \tfrac{1}{4}(1-\xi)(1+\eta)$$

ξ and η have the following values of the nodes:

node 1: $\xi = -1$ and $\eta = -1$

node 2: $\xi = 1$ and $\eta = -1$

node 3: $\xi = 1$ and $\eta = 1$

node 4: $\xi = -1$ and $\eta = 1$

The small deflection matrix for the in-plane plate, in global co-ordinates, is given by

$$[k^0] = \iint [B]^{\mathrm{T}}[D][B]\,\mathrm{d}x\,\mathrm{d}y$$

$$= \int_{-1}^{1}\int_{-1}^{1} [B]^{\mathrm{T}}[D][B]\,\det|J|\,\mathrm{d}\xi\,\mathrm{d}\eta \qquad (4.23)$$

where

$$[B] = [B_1 \;\; B_2 \;\; B_3 \;\; B_4]$$

and

$$[B_i] = \begin{bmatrix} \partial N_i/\partial x^0 & 0 \\ 0 & \partial N_i/\partial y^0 \\ \partial N_i/\partial y^0 & \partial N_i/\partial x^0 \end{bmatrix}$$

$$\begin{Bmatrix} \partial N_i/\partial x^0 \\ \partial N_i/\partial y^0 \end{Bmatrix} = [J^{-1}]\begin{Bmatrix} \partial N_i/\partial \xi \\ \partial N_i/\partial \eta \end{Bmatrix}$$

$$\det|J| = \frac{\partial(x^0, y^0)}{\partial(\xi, \eta)} = \begin{vmatrix} \partial x^0/\partial\xi & \partial y^0/\partial\xi \\ \partial x^0/\partial\eta & \partial y^0/\partial\eta \end{vmatrix}$$

$$|J| = \text{a Jacobian} = \begin{bmatrix} \partial N_1/\partial\xi & \partial N_2/\partial\xi \rightarrow \partial N_4/\partial\xi \\ \partial N_1/\partial\eta & \partial N_2/\partial\eta \rightarrow \partial N_4/\partial\eta \end{bmatrix}\begin{bmatrix} x_1^0 & y_1^0 \\ x_2^0 & y_2^0 \\ \downarrow & \downarrow \\ x_4^0 & y_4^0 \end{bmatrix}$$

$$[D] = E'\begin{bmatrix} 1 & \mu & 0 \\ \mu & 1 & 0 \\ 0 & 0 & \gamma \end{bmatrix} \qquad (4.24)$$

4.10.1. Geometrical Stiffness Matrix

To obtain $[k_G]$, it will be necessary to assume a value for w, the deflection perpendicular to the plane of the plate:

$$w = [N_1 \;\; N_2 \;\; N_3 \;\; N_4]\begin{Bmatrix} w_1 \\ w_2 \\ w_3 \\ w_4 \end{Bmatrix}$$

$$\begin{Bmatrix} \partial w/\partial x^0 \\ \partial w/\partial y^0 \end{Bmatrix} = \begin{bmatrix} \partial N_1/\partial x^0 & \partial N_2/\partial x^0 & \partial N_3/\partial x^0 & \partial N_4/\partial x^0 \\ \partial N_1/\partial y^0 & \partial N_2/\partial y^0 & \partial N_3/\partial y^0 & \partial N_4/\partial y^0 \end{bmatrix} \begin{Bmatrix} w_1 \\ w_2 \\ w_3 \\ w_4 \end{Bmatrix} \tag{4.25}$$

$$= [G] \begin{Bmatrix} w_1 \\ \downarrow \\ w_4 \end{Bmatrix}$$

$$[k_G] = \iint [G]^{\mathrm{T}}[\sigma][G]\, \mathrm{d}x^0\, \mathrm{d}y^0$$

$$[k_G] = \int_{-1}^{1}\int_{-1}^{1} [G]^{\mathrm{T}}[\sigma][G] \det|J|\, \mathrm{d}\xi\, \mathrm{d}\eta \tag{4.26}$$

where $[\sigma]$ is as defined in section 4.8.1.

4.10.2. The Shell Element

This can be obtained by considering the quadrilateral in local co-ordinates, as shown in Fig. 4.14. Putting $x_i^0 = y_i^0 = y_j^0 = 0$, and transforming other dimensions, etc., to local co-ordinates (see sections 4.10 and 4.10.1) the elemental stiffness matrix $[k]$ can be found.

To obtain the 'flat shell' element, it will now be necessary to

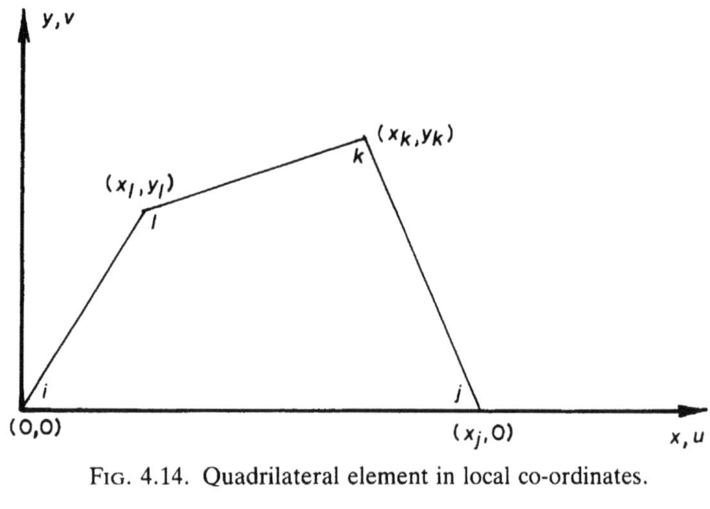

FIG. 4.14. Quadrilateral element in local co-ordinates.

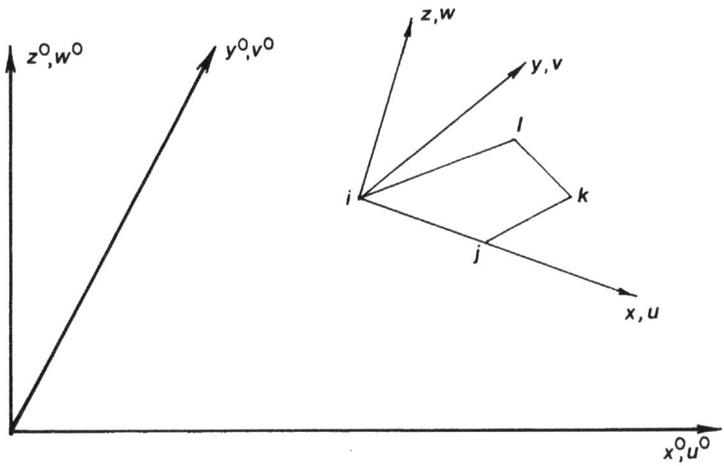

FIG. 4.15. 'Flat shell' quadrilateral element.

consider the in-plane plate of Fig. 4.15 in terms of the three-dimensional global system of Fig. 4.15.

The stiffness matrix, in global co-ordinates, is given by:

$$[k^0] = [\Xi]^T[k]_{12\times 12}[\Xi] \tag{4.27}$$

$$[\Xi] = \left[\begin{array}{c|c|c|c} \zeta & 0_3 & 0_3 & 0_3 \\ \hline 0_3 & \zeta & 0_3 & 0_3 \\ \hline 0_3 & 0_3 & \zeta & 0_3 \\ \hline 0_3 & 0_3 & 0_3 & \zeta \end{array}\right] \tag{4.28}$$

$[\zeta]$ is given by eqn (4.18), and $[k]_{12\times 12}$ is the expanded version of the elemental stiffness matrix of the in-plane quadrilateral, in local co-ordinates, where an additional four columns and rows of zeros have been allowed for the w displacements.

Similarly, the geometrical stiffness matrix for the 'flat shell' quadrilateral is given by:

$$[k_G^0] = [\Xi]^T[k_G]_{12\times 12}[\Xi] \tag{4.29}$$

where $[k_G]_{12\times 12}$ allows for eight columns and rows of zeros, corresponding to the u and v displacements.

4.11. EIGHT-NODE QUADRILATERAL ELEMENT

This element is a three-dimensional extension of the in-plane isoparametric quadrilateral element of Ergatoudis *et al.*,[15] and is shown in Fig. 4.16. The assumed displacement functions for this element are:

$$\begin{aligned} u^0 &= \alpha_1 + \alpha_2 x^0 + \alpha_3 y^0 + \alpha_4 x^0 y^0 + \alpha_5 x^{02} + \alpha_6 y^{02} \\ &\quad + \alpha_7 x^{02} y^0 + \alpha_8 y^{02} x^0 \\ v^0 &= \alpha_9 + \alpha_{10} x^0 + \alpha_{11} y^0 + \alpha_{12} x^0 y^0 + \alpha_{13} x^{02} + \alpha_{14} y^{02} \\ &\quad + \alpha_{15} x^{02} y^0 + \alpha_{16} y^{02} x^0 \\ w^0 &= \alpha_{17} + \alpha_{18} x^0 + \alpha_{19} y^0 + \alpha_{20} x^0 y^0 + \alpha_{21} x^{02} + \alpha_{22} y^{02} \\ &\quad + \alpha_{23} x^{02} y^0 + \alpha_{24} y^{02} x^0 \end{aligned}$$

which can be put in matrix form as:

$$\begin{Bmatrix} u^0 \\ v^0 \\ w^0 \end{Bmatrix} = \begin{bmatrix} N_1 & 0 & 0 & N_2 & 0 & 0 & N_3 & & \rightarrow & N_8 & 0 & 0 \\ 0 & N_1 & 0 & 0 & N_2 & 0 & 0 & N_3 & \rightarrow & 0 & N_8 & 0 \\ 0 & 0 & N_1 & 0 & 0 & N_2 & 0 & 0 & \rightarrow & 0 & 0 & N_8 \end{bmatrix} \begin{Bmatrix} u_1^0 \\ v_1^0 \\ w_1^0 \\ \downarrow \\ v_8^0 \\ w_8^0 \end{Bmatrix} \tag{4.30}$$

where $[N]$ is a matrix of shape functions, with

$$\begin{aligned} N_1 &= -\tfrac{1}{4}(1-\xi)(1-\eta)(\xi+\eta+1) \\ N_2 &= \tfrac{1}{2}(1-\xi^2)(1-\eta) \\ N_3 &= \tfrac{1}{4}(1+\xi)(1-\eta)(\xi-\eta-1) \\ N_4 &= \tfrac{1}{2}(1-\eta^2)(1+\xi) \\ N_5 &= \tfrac{1}{4}(1+\xi)(1+\eta)(\xi+\eta-1) \\ N_6 &= \tfrac{1}{2}(1-\xi^2)(1+\eta) \\ N_7 &= -\tfrac{1}{4}(1-\xi)(1+\eta)(\xi-\eta+1) \\ N_8 &= \tfrac{1}{2}(1-\eta^2)(1-\xi) \end{aligned}$$

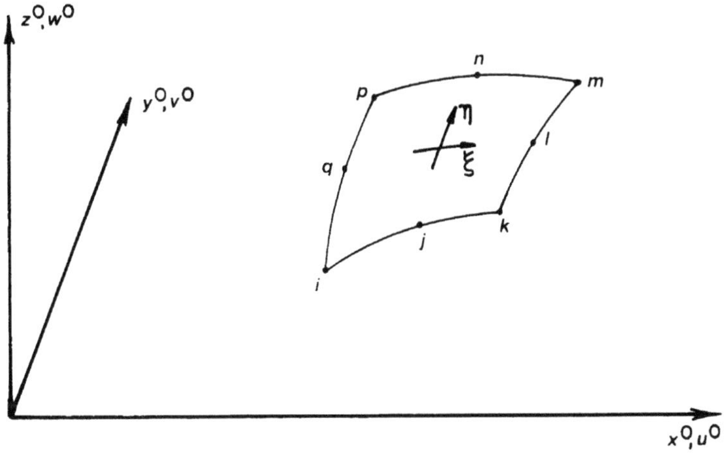

FIG. 4.16. Eight-node isoparametric element.

The curvilinear co-ordinates ξ and η point from the nodes q to l and from j to n, respectively, and the nodal values of ξ and η are as follows:

Node i	*Node j*	*Node k*
$\xi=-1,\ \eta=-1$	$\xi=0,\ \eta=-1$	$\xi=1,\ \eta=-1$
Node l	*Node m*	*Node n*
$\xi=1,\ \eta=0$	$\xi=1,\ \eta=1$	$\xi=0,\ \eta=1$
Node p	*Node q*	
$\xi=-1,\ \eta=1$	$\xi=-1,\ \eta=0$	

The elemental stiffness matrix for this curved membrane element is given by:

$$[k^0]_{24\times 24}=\int [B]^{\mathrm{T}}[D][B]\,\mathrm{d}x\,\mathrm{d}y \tag{4.31}$$

where x and y are local co-ordinates. Now

$$\begin{Bmatrix} u^0 \\ v^0 \\ w^0 \end{Bmatrix}=\sum N_i(\xi,\eta)\begin{Bmatrix} u_i^0 \\ v_i^0 \\ w_i^0 \end{Bmatrix}$$

and

$$\begin{Bmatrix} x^0 \\ y^0 \\ z^0 \\ t \end{Bmatrix} = \sum N_i(\xi, \eta) \begin{Bmatrix} x_i^0 \\ y_i^0 \\ z_i^0 \\ t_i \end{Bmatrix}$$

where t is wall thickness of the membrane. Furthermore

$$\begin{Bmatrix} \sigma_x \\ \sigma_y \\ \tau_{xy} \end{Bmatrix} = [D] \begin{Bmatrix} \dfrac{\partial u}{\partial x} \\ \dfrac{\partial v}{\partial y} \\ \dfrac{\partial u}{\partial y} + \dfrac{\partial v}{\partial x} \end{Bmatrix} = [D][B] \begin{Bmatrix} u_1^0 \\ v_1^0 \\ w_1^0 \\ u_2^0 \\ \downarrow \\ w_8^0 \end{Bmatrix} \tag{4.32}$$

where σ_x and σ_y are the direct stresses in the local x and y directions, respectively, and τ_{xy} is the shear stress in the x–y plane.

It can be seen from eqns (4.23) and (4.26) that, whereas the integration of $[B]$ is in terms of local co-ordinates, the matrix of shape functions $[N]$ is in terms of global co-ordinates. Thus, to obtain $[B]$, it will be necessary to resort to some elegant mathematics, as described by Irons and Ahmad.[8] Let $\boldsymbol{\xi}$ and $\boldsymbol{\eta}$ be base vectors, where

$$\begin{aligned} \boldsymbol{\xi} &= \lfloor \partial x^0/\partial \xi \;\; \partial y^0/\partial \xi \;\; \partial z^0/\partial \xi \rfloor \\ &= \lfloor \partial N_1/\partial \xi \;\; \partial N_2/\partial \xi \;\; \partial N_3/\partial \xi \rightarrow \partial N_8/\partial \xi \rfloor \begin{bmatrix} x_1^0 & y_1^0 & z_1^0 \\ x_2^0 & y_2^0 & z_2^0 \\ \downarrow & \downarrow & \downarrow \\ x_8^0 & y_8^0 & z_8^0 \end{bmatrix} \end{aligned} \tag{4.33}$$

and

$$\begin{aligned} \boldsymbol{\eta} &= \lfloor \partial x^0/\partial \eta \;\; \partial y^0/\partial \eta \;\; \partial z^0/\partial \eta \rfloor \\ &= \lfloor \partial N_1/\partial \eta \;\; \partial N_2/\partial \eta \;\; \partial N_3/\partial \eta \rightarrow \partial N_8/\partial \eta \rfloor \begin{bmatrix} x_1^0 & y_1^0 & z_1^0 \\ x_2^0 & y_2^0 & z_2^0 \\ \downarrow & \downarrow & \downarrow \\ x_8^0 & y_8^0 & z_8^0 \end{bmatrix} \end{aligned} \tag{4.34}$$

If $\hat{\mathbf{x}}$ is a unit vector in the local x direction, then

$$\hat{\mathbf{x}} = \boldsymbol{\xi}/|\boldsymbol{\xi}|$$

Irons and Ahmad[8] state that

$$\mathbf{y} = \boldsymbol{\eta} - \hat{\mathbf{x}} \cdot \boldsymbol{\eta}\hat{\mathbf{x}}$$

so that

$$\hat{\mathbf{y}} = \mathbf{y}/|\mathbf{y}|$$

and is the unit vector in the local y direction. Hence

$$\begin{Bmatrix} \dfrac{\partial}{\partial \xi} \\ \dfrac{\partial}{\partial \eta} \end{Bmatrix} = \begin{bmatrix} \boldsymbol{\xi} \cdot \hat{\mathbf{x}} & \boldsymbol{\xi} \cdot \hat{\mathbf{y}} \\ \boldsymbol{\eta} \cdot \hat{\mathbf{x}} & \boldsymbol{\eta} \cdot \hat{\mathbf{y}} \end{bmatrix} \begin{Bmatrix} \dfrac{\partial}{\partial x} \\ \dfrac{\partial}{\partial y} \end{Bmatrix} \tag{4.35}$$

or

$$[J] = \begin{bmatrix} \boldsymbol{\xi} \cdot \hat{\mathbf{x}} & \boldsymbol{\xi} \cdot \hat{\mathbf{y}} \\ \boldsymbol{\eta} \cdot \hat{\mathbf{x}} & \boldsymbol{\eta} \cdot \hat{\mathbf{y}} \end{bmatrix} \tag{4.36}$$

and

$$\begin{Bmatrix} \dfrac{\partial}{\partial x} \\ \dfrac{\partial}{\partial y} \end{Bmatrix} = [J]^{-1} \begin{Bmatrix} \dfrac{\partial}{\partial \xi} \\ \dfrac{\partial}{\partial \eta} \end{Bmatrix} \tag{4.37}$$

Irons and Ahmad[8] show that:

$$\begin{aligned} \frac{\partial u}{\partial x} &= \hat{\mathbf{x}}^{\mathrm{T}} \frac{\partial \lfloor N \rfloor}{\partial x} \{u_i^0\} \\ &= \hat{\mathbf{x}}^{\mathrm{T}} \left(\frac{\partial \lfloor N \rfloor}{\partial \xi} \frac{\partial \xi}{\partial x} + \frac{\partial \lfloor N \rfloor}{\partial \eta} \frac{\partial \eta}{\partial x} \right) \{u_i^0\} \\ \frac{\partial v}{\partial x} &= \hat{\mathbf{y}}^{\mathrm{T}} \left(\frac{\partial \lfloor N \rfloor}{\partial \xi} \frac{\partial \xi}{\partial x} + \frac{\partial \lfloor N \rfloor}{\partial \eta} \frac{\partial \eta}{\partial x} \right) \{u_i^0\} \\ \frac{\partial u}{\partial y} &= \hat{\mathbf{x}}^{\mathrm{T}} \left(\frac{\partial \lfloor N \rfloor}{\partial \eta} \frac{\partial \eta}{\partial y} + \frac{\partial \lfloor N \rfloor}{\partial \eta} \frac{\partial \eta}{\partial y} \right) \{u_i^0\} \end{aligned}$$

and

$$\frac{\partial v}{\partial y} = \hat{\mathbf{y}}^{\mathrm{T}} \left(\frac{\partial \lfloor N \rfloor}{\partial \eta} \frac{\partial \eta}{\partial y} + \frac{\partial \lfloor N \rfloor}{\partial \eta} \frac{\partial \eta}{\partial y} \right) \{u_i^0\}$$

From which $[B]$ can be obtained and, hence, $[k^0]_{24 \times 24}$.

It should be noted that $[B]$ is of order 3×24 and $\lfloor u_i^0 \rfloor = \lfloor u_i^0 \; v_i^0 \; w_i^0 \; u_j^0 \rightarrow v_q^0 w_q^0 \rfloor_{1 \times 24}$.

4.12. NUMERICAL INTEGRATION

For precise numerical integration, it is necessary in eqn (4.23) to use three Gauss points in the ξ direction and three points in the η direction per element. Similarly, as the powers of ξ and η are of higher order in eqn (4.30) than they are with eqn (4.21), it will be necessary in eqn (4.23) to use four Gauss points in the ξ direction and four Gauss points in the η direction per element.

For computational efficiency, reduced integration may be adopted. Stolarski and Belytschko[16] state that increased membrane stiffness, due to an excessive contribution of membrane effects, can be alleviated by reduced integration in this direction. This phenomenon of increased membrane stiffness, which results in numerical instability, is called *locking*.

4.13. PRACTICAL EXAMPLES

Detailed publications of practical examples showing the analysis of membrane structures are difficult to obtain. In this chapter, two examples will be given, and they cover the two major types of membrane structure, namely, air-supported structures and water-supported structures.

Newman *et al.*[1] carried out computational and experimental analysis on a thin-walled hemispherical dome, and also on two oblate domes of spherical shape with aspect ratios of 0·25 and 0·37, where

$$\text{aspect ratio (A.R.)} = \frac{\text{dome height}}{\text{base diameter}}$$

Newman *et al.* used a conventional approach, and adopted a four-node curved quadrilateral shell element.

A less conventional approach was adopted by Harrison[6] on the behaviour of inflatable dams, where he used one-dimensional elements. They appeared to be very effective and particularly efficient on computation.

Details of these two investigations will now be given.

4.14. STRESSES IN AIR-INFLATED SPHERICAL BUILDINGS

Newman *et al.* tested three model domes, of aspect ratios 0·25, 0·37 and 0·5, in a wind tunnel. The models had the dimensions shown in

TABLE 4.1
Membrane Models of Newman *et al.*

Model	*Base diameter (m)*	*Dome height (m)*	*A.R.*
1	0·463	0·116	0·25
2	0·455	0·168	0·37
3	0·457	0·229	0·5

Table 4.1. The material of construction of the membrane domes was impervious and practically unstretchable, and it resembled the material normally used for spinnaker-sails.

Newman *et al.* showed that if the minimum internal pressure on the upwind side of the domes fell to $180 \times (\text{A.R.}) \times \tau_W$, there was a possibility of dome buckling (A.R. is the aspect ratio; and τ_W is the skin friction at the ground $= \rho V_\tau^2$, where V_τ is the skin friction velocity and ρ is the density of air).

The computer analysis of Newman *et al.* adopted the four-node curved quadrilateral shell element of Bathe *et al.*,[17] together with the aid of their SAP IV finite element program.

In order to facilitate the presentation of their results, Newman *et al.* discretised the spherical domes with lines of co-latitude (θ = constant),

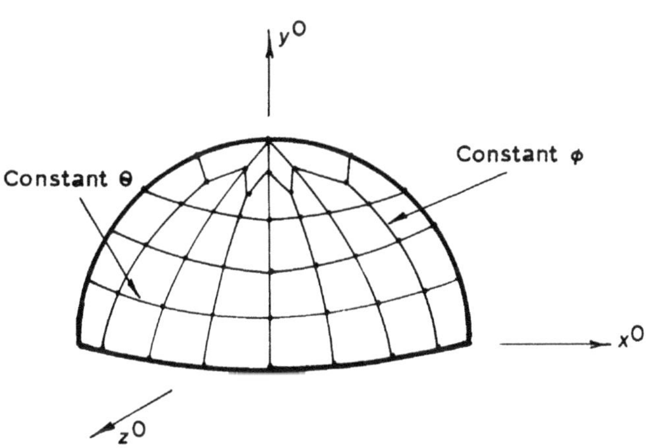

FIG. 4.17. Quadrilateral shell elements used to describe the domes.

together with lines of longitude (ϕ = constant), as shown in Fig. 4.17. To satisfy numerical precision, Newman *et al.* spaced the ϕ curves at 10 degrees apart and the θ curves at 3·5 degrees apart.

In order to ensure that the mathematical model was inextensible, Newman *et al.* assumed an elastic modulus of $10^{40}\,\mathrm{N\,m^{-2}}$, together with a wall thickness to base diameter ratio of $0{\cdot}22 \times 10^{-3}$. Newman *et al.* discovered that if the wall thickness to base diameter ratio were made equal to 0·022, the membrane equation was violated. Furthermore, they found that if the membrane was very thin (i.e. the wall thickness to base diameter ratio $= 0{\cdot}22 \times 10^{-10}$), large deflections occurred and the geometry of the membrane was significantly altered. They also found that, in changing Poisson's ratio from 0 to 0·3, there were no significant changes in tension. Newman *et al.* found that, in predicting membrane tensions, the stresses were very much independent of the material constants.

Newman *et al.* have presented the variation of tension with ϕ for various values of θ, as shown in Fig. 4.18. These figures also show a single broken curve, plotted against θ for $\phi = 0$. These single broken curves show that buckling occurs at $\phi = 0$ or on the windward side of the plane of symmetry of each model, where $C_{T\tau}$ is the principal membrane tension $(= \rho V_\tau^2 c/2)$, θ_g is the angle of the membrane at ground level, V_τ is the skin friction velocity $(= \tau_W/\rho)$, τ_W is the skin friction at the ground, ρ is the density of air, and c is the base diameter of the dome.

From their studies, Newman *et al.* have given the following empirical expression to determine the buckling pressures for spherical domes:

$$[(P_i - P_\infty)/(\tfrac{1}{2}\rho V_\tau^2)]/(\mathrm{A.R.}) = 360\,(\pm 1\%) \tag{4.38}$$

where P_i is the internal pressure, and P_∞ is the freestream static pressure.

Newman *et al.* have also computerised the variation of maximum principal tension against θ, for various values of ϕ, when buckling is just avoided, as shown in Fig. 4.19.

In their paper, Newman *et al.* have also discussed many other aspects of inflated buildings under air pressure. They state, for example, that the main method of failure of such structures usually occurs at the top of the structure, where tearing of the material takes place.

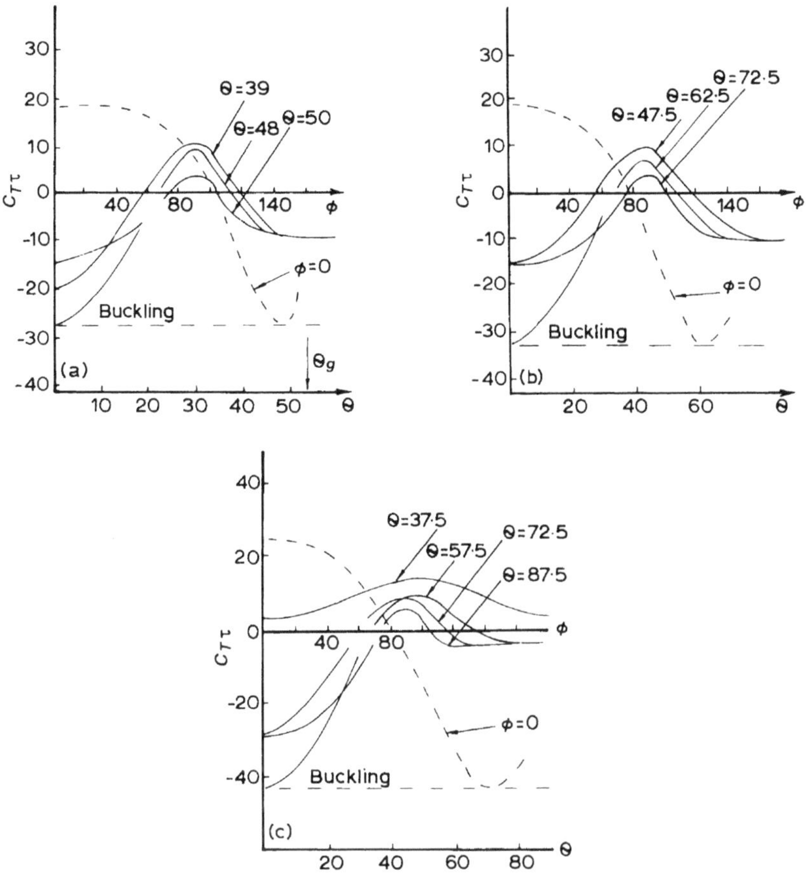

FIG. 4.18. Variation of minimum principal tension with ϕ and θ: (a) A.R. = 0·25; (b) A.R. = 0·37; (c) A.R. = 0·50. (θ and ϕ are measured in degrees.)

4.15. STRESSES IN WATER-SUPPORTED STRUCTURES

An example of the analysis of a water-supported membrane is that adopted by Harrison,[6] when he analysed membrane dams as high as 6·1 m.

As the membrane dam is of cylindrical form and its deformations are very large, Harrison analysed the cross-section of the dam, treating

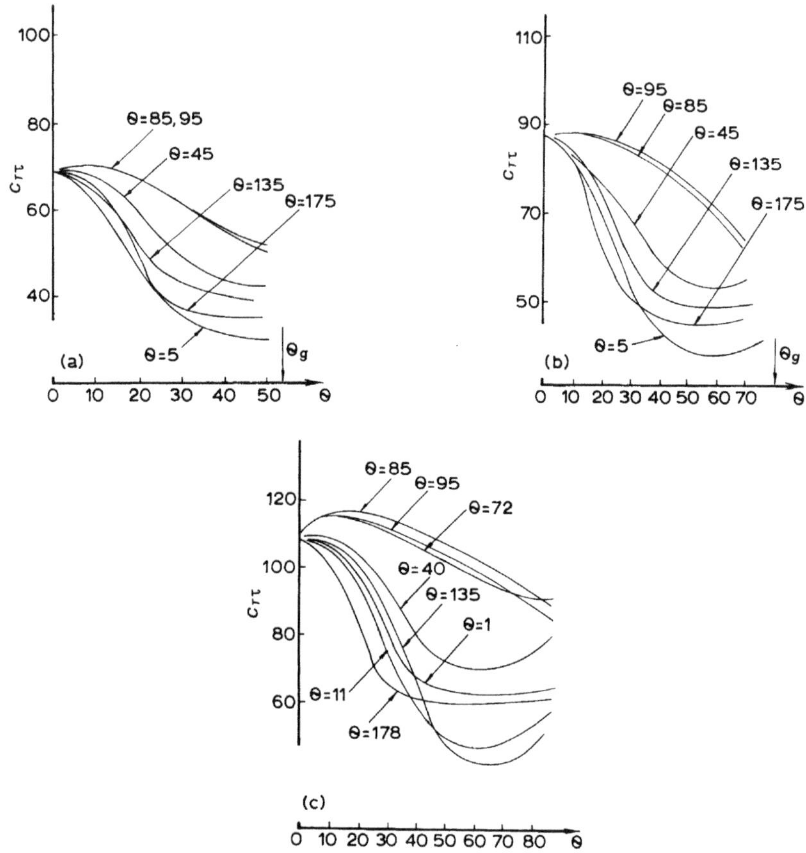

FIG. 4.19. Variation of maximum principal tension with θ, when buckling is just avoided: (a) A.R. = 0·25; (b) A.R. = 0·37; (c) A.R. = 0·50. (θ is measured in degrees.)

it as a two-dimensional structure composed of several one-dimensional elements. As the cross-section of the membrane was clamped at its two ends, (Fig. 4.20), the co-ordinates of the first and last nodes were known. Assuming there were n elements, then there were $n + 1$ nodes, and the solution required to determine n element tensions, and $2n - 2$ co-ordinates and a total of $3n - 2$ unknowns.

Harrison shows, from equilibrium considerations, that $2n - 2$ equations can be obtained, and that a further n equations can be obtained

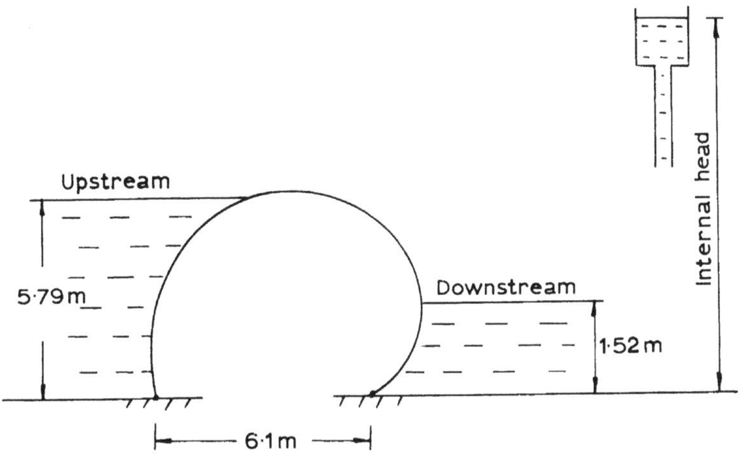

FIG. 4.20. A water-inflated membrane dam. Curved perimeter = 18·29 m; thickness = 7·62 mm; elastic modulus = $2{\cdot}07 \times 10^9$ Pa.

from compatibility considerations, thus ensuring that the required number of equations can be obtained.

Harrison's analysis assumed the membrane tension in element 1, together with its slope, so that the co-ordinates of node 2 could be readily determined. He then assumed that the water and/or air pressures on element 1 acted at node 2 and, from equilibrium considerations, determined the horizontal and vertical components of membrane tension in element 2. Hence he was able to calculate the preliminary values for the co-ordinates of node 3, from these membrane tensile components. The process was then continued until the co-ordinates of the last node were computed. If these computed co-ordinates did not agree with their actual fixed values, then the process was repeated with another initial value for tension and slope of the first element.

To demonstrate the method, Harrison considered the dam of Fig. 4.20, which had an unstretched perimeter of 18·29 m, with its base fixed at two points 6·1 m apart. The dam was assumed to contain a head of water of 5·79 m on its upstream side, and it had a head of water of 1·52 m at its downstream end.

Figure 4.21 shows the relationship between internal water head and dam height, where it can be seen that, to maintain stability, an internal head of 10·06 m was required.

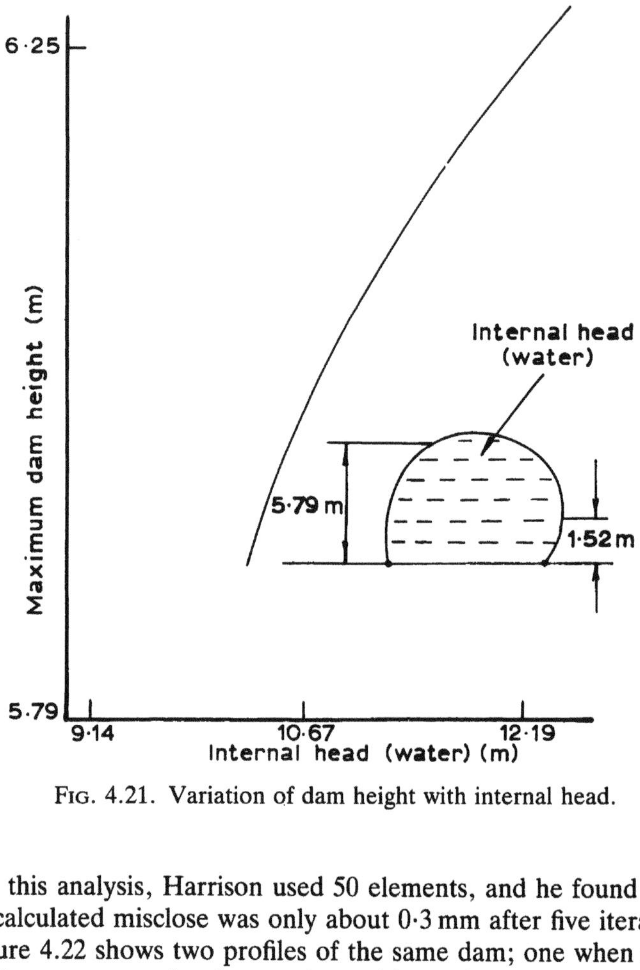

FIG. 4.21. Variation of dam height with internal head.

For this analysis, Harrison used 50 elements, and he found that the total calculated misclose was only about 0·3 mm after five iterations.

Figure 4.22 shows two profiles of the same dam; one when resisting a 5·79 m upstream head, together with an internal water head of 10·7 m, and the other with the same internal water head, but with values of both upstream and downstream heads of 1·52 m. In this latter case, Harrison found that the membrane hoop tension was 235·24 kN m^{-1}, but this reduced to 219·7 kN m^{-1} when this upstream head rose to 5·79 m.

Harrison found that the membrane dam behaved unusually because its hoop membrane tension decreased with a rising upstream head (Fig. 4.23).

Harrison also plotted the variation of dam height with a rising

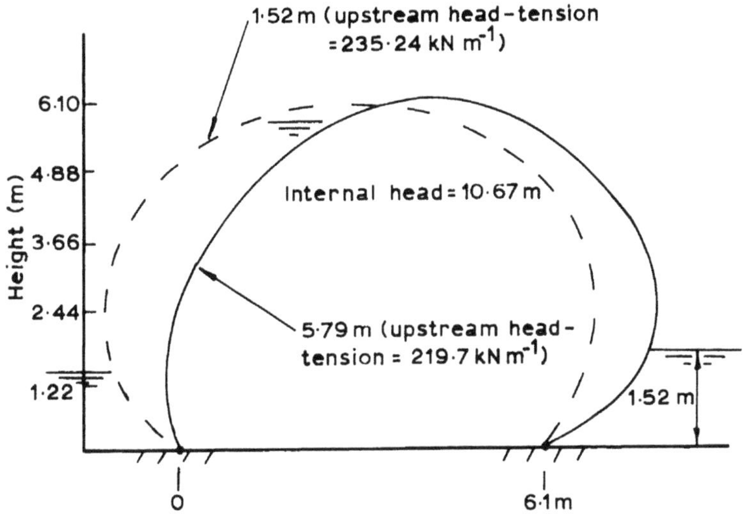

FIG. 4.22. Profiles of dam with different upstream heads.

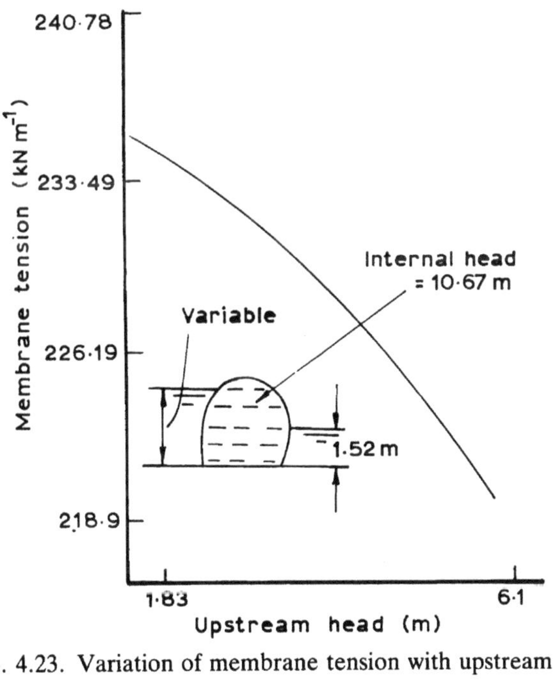

FIG. 4.23. Variation of membrane tension with upstream head.

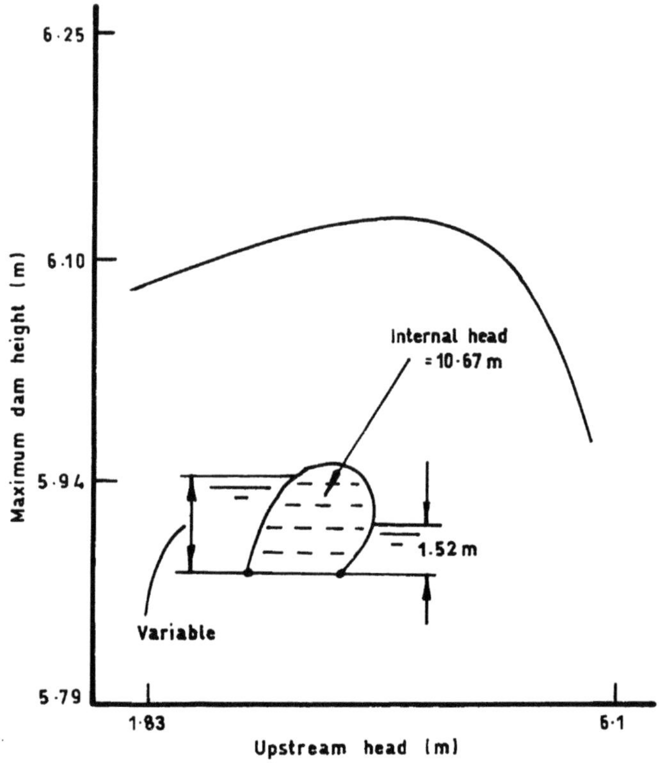

FIG. 4.24. Variation of dam height with upstream head.

upstream head, (Fig. 4.24), where it can be seen that the dam height increases before it falls.

Harrison also investigated the same dam under air pressure, where he plotted the variation of maximum dam height with increasing upstream head for two different pressures, (Fig. 4.25).

In these two cases, Harrison found that the dam height increased with increasing upstream head, before it suddenly fell.

Harrison also plotted the variation of membrane tension against increasing upstream head for the two different internal air pressures (Fig. 4.26).

Finally, he plotted the variation of enclosed cross-sectional area of the membrane dam with increasing upstream head, for an internal air pressure of 27·58 kPa (Fig. 4:27).

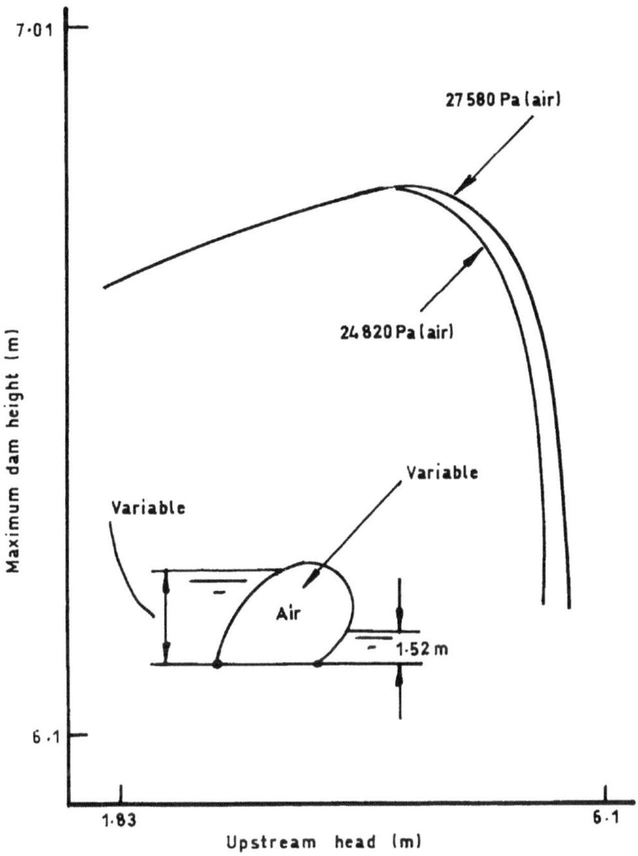

FIG. 4.25. Variation of dam height with upstream head and under internal air pressure.

Harrison concludes that a water-inflated membrane is safer than an air-inflated membrane, as the latter is in danger of explosive failure and will require a larger factor of safety.

4.16. CONCLUSIONS

The chapter has shown that the finite element method is a very powerful tool for analysing membrane structures, varying from cables under self-weight to water-inflated dams and air-inflated buildings.

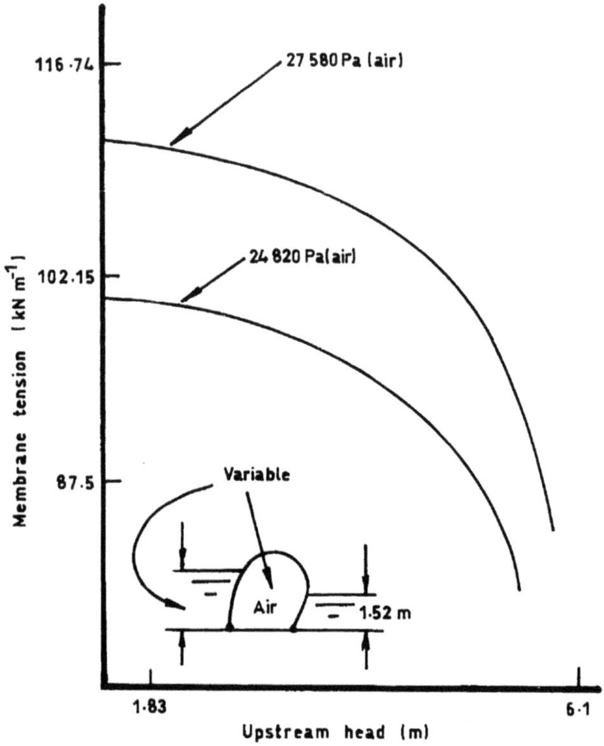

FIG. 4.26. Variation of membrane tension with upstream head and under internal air pressure.

For curved membranes, if the deflections are not large, there is no need to use the geometrical stiffness matrix in addition to the small deflection stiffness matrix, as a quite efficient analysis can be based on small deflection linear elastic theory.

Axisymmetric membrane elements have not been discussed in this chapter, but they could be successfully applied to the design of axisymmetric structures, varying from storage tanks[17] to airships.

Finally, it can be concluded, that with the growing interest in leisure activities, particularly in cold climates, the use of large membrane structures will continue to gain popularity, as for example the 40 000 m^2 floor area membrane building recently constructed in Canada.

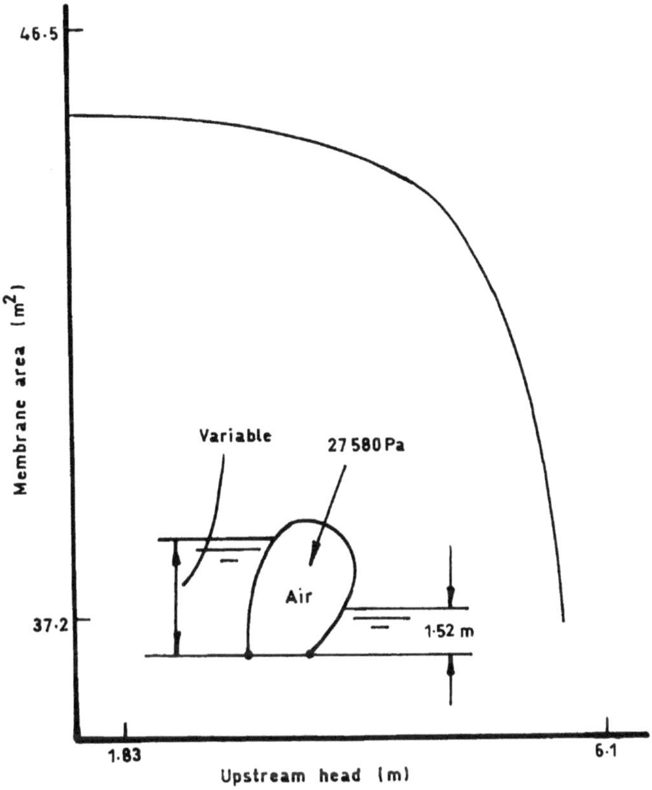

FIG. 4.27. Variation of membrane sectional area with upstream head and under an internal air pressure of 27·58 kPa.

ACKNOWLEDGEMENTS

The author would like to thank Mr Ken Mildren of the Portsmouth Polytechnic Library for his co-operation, and also Mrs Lesley Jenkinson for the care and devotion she showed in typing this chapter.

REFERENCES

1. NEWMAN, B. G., GANGULI, U. and SHRIVASTAVA, S. C., Flow over spherical inflated buildings, *J. Wind Engng Ind. Aerodyn.*, **17**, 1984, 305–27.

2. NEWBY, F., SUAN, R. H. and FELIX, J. S. (Cedric Price and Partners, Architects) *Air Structures: A Survey,* Department of the Environment, HMSO, London, 1917.
3. BIRD, W. W., Design manual for spherical radomes', *Report UB-909-D2,* Cornell Aeronautical Laboratory, 1965.
4. SPRINGFIELD, J. and SINORSKI, D., The air supported steel membrane roof at Dalhousie University, Halifax, Nova Scotia, *Can. Struct. Engng Conf. on Canadian Steel Construction,* Ontario, 1980, pp. 1–29.
5. SHEPHERD, E. M., MCKAY, F. A. and HODGENS, V. T., The Fabridam extension on Koombooloomba dam on the Tully Falls hydro-electric power project, *J. Instn Engrs, Aust.,* **41,** 1969, 1–7.
6. HARRISON, H. B., The analysis and behaviour of inflatable membrane dams under static loading, *Proc. I.C.E.,* **45,** 1970, 661–76.
7. ZIENKIEWICZ, O. C., *The Finite Element Method in Engineering Science,* 3rd edn., McGraw-Hill, New York, 1977.
8. IRONS, B. and AHMAD, S., *Techniques of Finite Elements,* Ellis Horwood, Chichester, 1980.
9. ROSS, C. T. F., *The Finite Element Method in Structural Mechanics,* Ellis Horwood, Chichester, 1985.
10. SACK, R. L., *Structural Analysis,* McGraw-Hill, New York, 1984.
11. TURNER, M. J., CLOUGH, R. W., MARTIN, H. C. and TOPP, L. J., Stiffness and Deflection Analysis of Complex Structures, *J. Aeronaut. Sci.,* **23,** 1956, 805–23.
12. TAIG, I. C., Structural analysis by the matrix displacement method, *Report No. SO17,* English Electric Aviation, 1961.
13. AHMAD, S., IRONS, B. M. and ZIENKIEWICZ, O. C., Curved thick shell and membrane elements with particular reference to axisymmetric problems, *Proc. Conf. on Matrix Methods in Structural Mechanics,* A.F.B., Ohio, AFFDL-TR-68-150, 1968, pp. 539–72.
14. PETYT, M., The application of finite element techniques to plate and shell problems, *Report No. 120,* ISVR, Univ. Southampton, Feb. 1965.
15. ERGATOUDIS, J. G., IRONS, B. M. and ZIENKIEWICZ, O. C., Curved isoparametric quadrilateral elements for finite element analysis, *Int. J. Solids Struct.,* **4,** 1968, 31–42.
16. STOLARSKI, H. and BELYTSCHKO, T., Shear/membrane locking in curved finite elements, *Proc. 4th Engng Mech. Conf. on Recent Advances,* New York, USA, 1983, vol. 2, pp. 830–33.
17. BATHE, K., WILSON, E. and PETERSON, F., A structural analysis program for static and dynamic response of linear systems, *Rep. EERC 73-11,* Univ. California, Berkeley, 1973.
18. ROTTER, N. L. and TREHAIR, N. S., Stress distribution in a fermentation tank, *Report CE24,* Instn Engrs, Australia, 1982, pp. 40–46.

Chapter 5

Axisymmetric Thin Shells

D. Hitchings
Department of Aeronautics, Imperial College of Science and Technology, London, UK

5.1. INTRODUCTION

Axisymmetric thin shells are common structural elements and are found in many areas of engineering. They have the advantages of being structurally efficient and relatively easy to fabricate. The cylinder is the most common form of the axisymmetric shell, but other examples are spheres, cones and toroids. Their use includes pressure vessels, cooling towers, wheels, tyres and turbine engine components. Their use spans all branches of engineering. They form efficient structures because they can be designed to carry loads by membrane action over most of their length. Problems arise where different shells are interconnected, where loads are applied, or where there is any other form of discontinuity. Local self-equilibrating stress systems are set up at the points of such discontinuities, and a major part of the analysis of axisymmetric shells consists of calculating these discontinuity stresses.

There are two categories of axisymmetric shell finite element, thin shells, where the stresses and strains through the wall thickness are assumed to be zero and thick shells, where this assumption is not made. Here, emphasis is placed upon the thin-shell formulation since, although the thick-shell description is more general, it does give rise to numerical conditioning problems as the wall thickness of the shell becomes thinner. A true thin-shell analysis, which assumes that thickness effects can be ignored, can be both more accurate and cheaper to use, as the ratio of radius to thickness increases. Chapter 14 of ref. 1 gives a good overview of the history and application of the finite element method to the analysis of thin shells. Probably the first development of such an element was by Grafton and Strome,[2] where the element was taken as a straight-sided conical frustrum. The loads were also assumed to be axisymmetric. The extension to non-axisymmetric loadings was given by Klein.[3] Higher order curved shell elements were developed by Chan and Firmin,[4] where they were applied to the analysis of a cooling tower. It was shown that good results could be obtained with a small number of high order elements, and also how axisymmetric shells could be used in non-linear large-deflection analysis. All of these references used the displacement method to formulate the shell element. Mixed formulations have also been developed, notably by Chan and Trbojevic.[5] Axisymmetric elements are very efficient for calculating mode shapes and frequencies of shells.[6] One comprehensive set of axisymmetric shell analysis computer programs is BOSOR4[7] and BOSOR5.[8] This is not strictly a finite element formulation, and is described as a 'finite-difference' element, but its application is very close to that employed by finite elements. They use a 7×7 element stiffness matrix corresponding to a constant strain, constant curvature change finite element. It is a non-conforming element in that the normal displacements and rotations across the element boundaries are not compatible, but a rapid convergence with increasing mesh density is claimed. The programs provide not only a linear static analysis of complex branched shells but also allow normal modes and buckling loads to be found, together with a moderately large deflection analysis. BOSOR5 also includes elastic–plastic and creep analysis. The programs allow considerable flexibility in the geometrical modelling, including layered material and shells with a non-axisymmetric geometry.

5.2. AXISYMMETRIC SHELL EQUATIONS

A typical frustrum of an axisymmetric shell is shown in Fig. 5.1. The shell has a principal radius of curvature R_ϕ, and a cylindrical radius, R. It is defined by a line, usually the mid-thickness, and it is assumed that normals to this line remain straight. The shell has three components of strain, the axial strain, ε_{SS}, the hoop strain, $\varepsilon_{\theta\theta}$, and

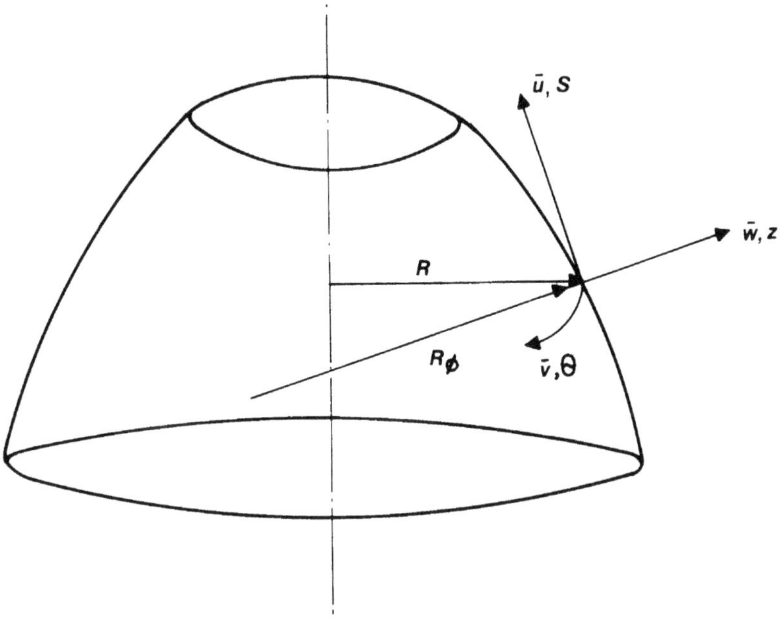

FIG. 5.1. Axisymmetric shell geometry and displacements.

the in-plane shear strain, $\varepsilon_{S\theta}$. It is assumed that the shell is thin so that the thickness components of stress and strain can be ignored. The three non-zero strain components can be expressed in terms of the three local displacements, $\bar{u}$ in the local S direction, $\bar{w}$ normal to the shell surface in the direction of the principal radius of curvature, R_ϕ, and $\bar{v}$ tangential to the shell circumference. The strain displacement relationships can be written as

$$\boldsymbol{\varepsilon} = \bar{\boldsymbol{\delta}}\bar{\mathbf{u}} = \bar{\boldsymbol{\delta}}_1\bar{\mathbf{u}} + z\bar{\boldsymbol{\delta}}_2\bar{\mathbf{u}}$$

where $\boldsymbol{\varepsilon}$ and $\bar{\mathbf{u}}$ are the column vectors

$$\boldsymbol{\varepsilon} = \{\varepsilon_{SS}\ \ \varepsilon_{\theta\theta}\ \ \varepsilon_{S\theta}\}$$
$$\bar{\mathbf{u}} = \{\bar{w}\ \ \bar{u}\ \ \bar{v}\}$$

The strain operator matrix, $\bar{\boldsymbol{\delta}}$, is composed of two parts; $\bar{\boldsymbol{\delta}}_1$, the membrane component which is constant through the thickness and $\bar{\boldsymbol{\delta}}_2$, the bending component which varies linearly through the wall thickness in the z direction:

$$\bar{\boldsymbol{\delta}}_1 = \begin{bmatrix} \dfrac{1}{R_\phi} & \dfrac{\delta}{\delta S} & 0 \\ \dfrac{\sin\phi}{R} & \dfrac{\cos\phi}{R} & \dfrac{1}{R}\dfrac{\delta}{\delta\theta} \\ 0 & \dfrac{1}{R}\dfrac{\delta}{\delta\theta} & \dfrac{\delta}{\delta S} - \dfrac{\cos\phi}{R} \end{bmatrix}$$

$$\bar{\boldsymbol{\delta}}_2 = \begin{bmatrix} -\dfrac{\delta^2}{\delta S^2} & \dfrac{1}{R_\phi}\dfrac{\delta}{\delta S} - \dfrac{1}{R_\phi^2}\dfrac{\delta R_\phi}{\delta S} & 0 \\ -\dfrac{1}{R^2}\dfrac{\delta^2}{\delta\theta^2} - \dfrac{\cos\phi}{R}\dfrac{\delta}{\delta S} & \dfrac{\cos\phi}{RR_\phi} & \dfrac{\sin\phi}{R^2}\dfrac{\delta}{\delta\theta} \\ -\dfrac{1}{R}\dfrac{\delta^2}{\delta\theta\delta S} + \dfrac{\cos\phi}{R^2}\dfrac{\delta}{\delta\theta} & \dfrac{1}{RR_\phi}\dfrac{\delta}{\delta\theta} & \dfrac{\sin\phi}{R}\dfrac{\delta}{\delta S} - \dfrac{\sin\phi\cos\phi}{R^2} \end{bmatrix}$$

In forming these strain–displacement relationships some assumptions must be made, and different authors quote slight variations to these operators.[9,10] In practice, the variations make little difference to the analysis and they only become significant if there are rapid variations of displacements around the shell circumference. The actual variation of the displacement around the circumference can be expressed as a Fourier series in the form:

$$\bar{w} = \sum_n w_{1n} \cos n\theta + \sum_n w_{2n} \sin n\theta$$

$$\bar{u} = \sum_n u_{1n} \cos n\theta + \sum_n u_{2n} \sin n\theta$$

$$\bar{v} = \sum_n v_{1n} \sin n\theta + \sum_n v_{2n} \cos n\theta$$

If the problem that is being analysed is linear, or it has a uniform geometry and material properties around the circumference, then it will be shown later that each term in the Fourier series can be solved as a separate problem. In many cases this simplifies the analysis considerably and is one of the reasons why an axisymmetric model is to be preferred to a general shell. Taking one term in the series, the displacements can be written as

$$\bar{w} = w \cos n\theta$$
$$\bar{u} = u \cos n\theta$$
$$\bar{v} = v \sin n\theta$$

and the strain displacement relationships then become

$$\boldsymbol{\varepsilon} = \mathbf{T}\boldsymbol{\delta}_1\mathbf{u} + \mathbf{T}\boldsymbol{\delta}_2\mathbf{u}$$

where

$$\mathbf{u} = \{w \;\; u \;\; v\}$$

$$\mathbf{T} = \begin{bmatrix} \cos n\theta & 0 & 0 \\ 0 & \cos n\theta & 0 \\ 0 & 0 & \sin n\theta \end{bmatrix}$$

$$\boldsymbol{\delta}_1 = \begin{bmatrix} \dfrac{1}{R_\phi} & \dfrac{\delta}{\delta S} & 0 \\ \dfrac{\sin\phi}{R} & \dfrac{\cos\phi}{R} & \dfrac{n}{R} \\ 0 & -\dfrac{n}{R} & \dfrac{\delta}{\delta S} - \dfrac{\cos\phi}{R} \end{bmatrix}$$

$$\boldsymbol{\delta}_2 = \begin{bmatrix} -\dfrac{\delta^2}{\delta S^2} & \dfrac{1}{R_\phi}\dfrac{\delta}{\delta S} - \dfrac{1}{R_\phi^2}\dfrac{\delta R_\phi}{\delta S} & 0 \\ \dfrac{n^2}{R^2} - \dfrac{\cos\phi}{R}\dfrac{\delta}{\delta S} & \dfrac{\cos\phi}{RR_\phi} & \dfrac{n\sin\phi}{R^2} \\ \dfrac{n}{R}\dfrac{\delta}{\delta S} - \dfrac{n\cos\phi}{R^2} & -\dfrac{n}{RR_\phi} & \dfrac{\sin\phi}{R}\dfrac{\delta}{\delta S} - \dfrac{\sin\phi\cos\phi}{R^2} \end{bmatrix}$$

These are used as the strain displacement relationships for the finite element analysis.

5.3. FINITE ELEMENT FORMULATION

Various forms of finite elements have been devised to solve the axisymmetric shell problem.[1–7] In all cases the element is based upon a shell frustrum, as is shown in Fig. 5.1. Such elements are unusual in that the nodes are nodal circles rather than nodal points. The geometry is defined in terms of a line on the cross-section of the shell. This need not be the centreline, and a variation of thickness along the line is allowed, with different values in thickness on each side of the line. The nodal loads can be defined in various ways. They can be total loads on the circumference, loads per unit length of the circumference or loads per radian. This latter description (as loads per radian) carries over directly to the non-axisymmetric loading case and will be used here.

The earliest form of axisymmetric shell elements assumed a conical frustrum, with straight sides.[1] Higher order elements can allow for a curved geometry. The nodal displacements must include displacements in all three directions and, since it is a shell element with bending, the rotation about the tangential axis must also be included. Some elements have been derived to include higher order derivatives as nodal freedoms[4] but if this is done then care must be taken in implementing the element. Typically, if the first derivative of the longitudinal displacement or the second derivative of the transverse displacement are used, then axial forces and bending moments are being matched across nodes. If the shell has any discontinuities at the node then these quantities will not be continuous, and matching them can lead to significant errors. In the element used here to illustrate the derivation of the stiffness matrices the three-noded, variable thickness element shown in Fig. 5.2 is used. There are four degrees of freedom

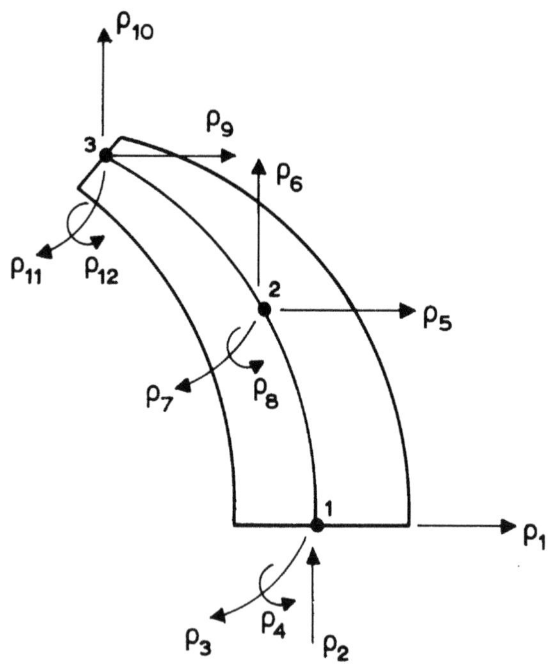

FIG. 5.2. Finite element geometry and displacements.

at each node, the three translations and the rotation about the circumference of the shell. This allows the axial and tangential displacements to vary quadratically along the shell, and the displacement normal to the shell surface can vary as a fifth-order polynomial.

The geometry of the cross-section line can vary parabolically, allowing the shell to be curved. The co-ordinates at the nodes are

$$\mathbf{x}_N = \{x_1 \;\; x_2 \;\; x_3\}$$
$$\mathbf{y}_N = \{y_1 \;\; y_2 \;\; y_3\}$$

and the thickness at the three nodes, normal to the line 123 are, to the left of this line,

$$\mathbf{t}_{L_N} = \{t_{L_1} \;\; t_{L_2} \;\; t_{L_3}\}$$

and to the right of the line

$$\mathbf{t}_{R_N} = \{t_{R_1} \;\; t_{R_2} \;\; t_{R_3}\}$$

Using Lagrange interpolation, the co-ordinates at any point are

$$x = \mathbf{v}\mathbf{x}_N, \qquad y = \mathbf{v}\mathbf{y}_N$$

where

$$\mathbf{v} = [v_1 \;\; v_2 \;\; v_3]$$
$$v_1 = -\tfrac{1}{2}\zeta(1-\zeta), \qquad v_2 = 1-\zeta^2, \qquad v_3 = \tfrac{1}{2}\zeta(1+\zeta)$$

and ζ is the non-dimensional co-ordinate along the nodal line. It has the value of -1 at node 1, 0 at node 2 and $+1$ at node 3. From these interpolations

$$\mathrm{d}x = \mathbf{v}_{,\zeta}\mathbf{x}_N \,\mathrm{d}\zeta, \qquad \mathrm{d}y = \mathbf{v}_{,\zeta}\mathbf{y}_N \,\mathrm{d}\zeta$$

where

$$\mathbf{v}_{,\zeta} = \mathrm{d}\mathbf{v}/\mathrm{d}\zeta = [(\zeta - \tfrac{1}{2}) \;\; -2\zeta \;\; (\zeta + \tfrac{1}{2})]$$

The increment of arc length at any point is

$$\mathrm{d}S = (\mathrm{d}_x^2 + \mathrm{d}_y^2)^{1/2} = \gamma \,\mathrm{d}\zeta$$
$$\gamma = [(\mathbf{v}_{,\zeta}\mathbf{x}_N)^2 + (\mathbf{v}_{,\zeta}\mathbf{y}_N)^2]^{1/2}$$

and the local co-ordinate directions are

$$\cos\phi = \mathrm{d}x/\mathrm{d}S = \frac{1}{\gamma}(\mathbf{v}_{,\zeta}\mathbf{x}_N)$$

$$\sin\phi = \mathrm{d}y/\mathrm{d}S = \frac{1}{\gamma}(\mathbf{v}_{,\zeta}\mathbf{y}_N)$$

The through-thickness co-ordinate at any point is

$$z = \tfrac{1}{2}(1-\eta)\mathbf{vt}_{\mathrm{L_N}} + \tfrac{1}{2}(1+\eta)\mathbf{vt}_{\mathrm{R_N}}$$
$$\mathrm{d}z = (-\tfrac{1}{2}\mathbf{vt}_{\mathrm{L_N}} + \tfrac{1}{2}\mathbf{vt}_{\mathrm{R_N}})\,\mathrm{d}\eta = \delta\,\mathrm{d}\eta$$

where η is again a non-dimensional co-ordinate that varies between -1 and $+1$ across the thickness. The cylindrical radius, R, at any point is the x co-ordinate, and the inverse of the principle radius of curvature is[1]

$$\frac{1}{R_\phi} = \frac{\mathrm{d}^2y/\mathrm{d}x^2}{[1+(\mathrm{d}y/\mathrm{d}x)^2]^{3/2}} = \left[\frac{\mathrm{d}^2y}{\mathrm{d}\xi^2}\frac{\mathrm{d}x}{\mathrm{d}\xi} - \frac{\mathrm{d}^2x}{\mathrm{d}\xi^2}\frac{\mathrm{d}y}{\mathrm{d}\xi}\right] \Big/ \left[\left(\frac{\mathrm{d}x}{\mathrm{d}\xi}\right)^2 + \left(\frac{\mathrm{d}y}{\mathrm{d}\xi}\right)^2\right]^{-1/2}$$

This defines the geometry of the shell.

The nodal displacements are

$$\boldsymbol{\rho} = \{\rho_1\ \rho_2\ \rho_3\ \rho_4\ \rho_5\ \rho_6\ \rho_7\ \rho_8\ \rho_9\ \rho_{10}\ \rho_{11}\ \rho_{12}\}$$

and these are used to interpolate for the displacements at any point in the form

$$\mathbf{u} = \begin{bmatrix} \bar{w} \\ \bar{u} \\ \bar{v} \end{bmatrix} = \begin{bmatrix} w_1 & 0 & 0 & w_2 & w_3 & 0 & 0 & w_4 & w_5 & 0 & 0 & w_6 \\ 0 & v_1 & 0 & 0 & 0 & v_2 & 0 & 0 & 0 & v_3 & 0 & 0 \\ 0 & 0 & v_1 & 0 & 0 & 0 & v_2 & 0 & 0 & 0 & v_3 & 0 \end{bmatrix} \mathbf{T}_{\mathrm{N}}\boldsymbol{\rho}$$
$$= \mathbf{w}\mathbf{T}_{\mathrm{N}}\boldsymbol{\rho}$$

where the terms v_1, v_2 and v_3 are exactly as for the geometrical interpolation and

$$w_1 = (4\xi^2 - 5\xi^3 - 2\xi^4 + 3\xi^5)/4$$
$$w_2 = \gamma_1(\xi^2 - \xi^3 - \xi^4 + \xi^5)/4$$
$$w_3 = (1 - 2\xi^2 + \xi^4)$$
$$w_4 = \gamma_2(\xi - 2\xi^3 + \xi^5)$$
$$w_5 = (4\xi^2 + 5\xi^3 - 2\xi^4 - 3\xi^5)/4$$
$$w_6 = \gamma_3(-\xi^2 - \xi^3 + \xi^4 - \xi^5)/4$$

where γ_i is the value of γ at the ith node.

The local nodal displacements can be transformed to global co-ordinates as

$$\mathbf{T}_{\mathrm{N}} = \begin{bmatrix} T_{\mathrm{N_1}} & 0 & 0 \\ 0 & T_{\mathrm{N_2}} & 0 \\ 0 & 0 & T_{\mathrm{N_3}} \end{bmatrix}$$

where

$$\mathbf{T}_{\mathrm{N}_i} = \begin{bmatrix} \sin\phi_i & -\cos\phi_i & 0 \\ \cos\phi_i & \sin\phi_i & 0 \\ 0 & 0 & 1 \end{bmatrix}$$

and ϕ_i is the local angle at node i. These are found from the geometry interpolation. In order to simplify the derivation of the element, some assumptions are now made. It will always be necessary to use relatively short elements in regions where the rates of change of strain are highest, so that it is possible to assume that the change in curvature of the element is small. This assumption has two consequences, first that the term involving the derivative of the principal radius of curvature in the axial bending strain can be assumed zero and, secondly, that the rate of change of the local angles, ϕ_i, around the shell can be assumed zero. With these assumptions the strain–displacement relationship becomes

$$\boldsymbol{\varepsilon} = (\boldsymbol{\alpha}_1 + z\boldsymbol{\alpha}_2)\mathbf{T}_{\mathrm{N}}\boldsymbol{\rho} = \boldsymbol{\alpha}\mathbf{T}_{\mathrm{N}}\boldsymbol{\rho}$$

$$\boldsymbol{\alpha}_1 = \boldsymbol{\delta}_1\mathbf{w}, \qquad \boldsymbol{\alpha}_2 = \boldsymbol{\delta}_2\mathbf{w}$$

The strain–displacement matrices, $\boldsymbol{\alpha}_1$ for the membrane and $\boldsymbol{\alpha}_2$ for bending, can be found by formal differentiation of the interpolation matrix, $\mathbf{w}$. To do this note that

$$\frac{\delta}{\delta S} = \frac{1}{\gamma}\frac{\delta}{\delta \zeta}$$

Following standard finite element theory, the element stiffness matrix can then be written as

$$\mathbf{k} = \mathbf{T}_{\mathrm{N}}^{\mathrm{t}} \int_V \boldsymbol{\alpha}^{\mathrm{t}}\boldsymbol{\kappa}\boldsymbol{\alpha}\,\mathrm{d}V\,\mathbf{T}_{\mathrm{N}}$$

$$= c\mathbf{T}_{\mathrm{N}}^{\mathrm{t}} \int_{-1}^{1}\int_{-1}^{1} (\boldsymbol{\alpha}_1^{\mathrm{t}} + z\boldsymbol{\alpha}_2^{\mathrm{t}})\boldsymbol{\kappa}(\boldsymbol{\alpha}_1 + z\boldsymbol{\alpha}_2)\gamma\delta R\,\mathrm{d}\eta\,\mathrm{d}\zeta\,\mathbf{T}_{\mathrm{N}}$$

where $\boldsymbol{\kappa}$ is the material stiffness matrix. For an isotropic homogeneous material this is

$$\boldsymbol{\kappa} = \frac{E}{(1-\nu^2)}\begin{bmatrix} 1 & \nu & 0 \\ \nu & 1 & 0 \\ 0 & 0 & \dfrac{1-\nu}{2} \end{bmatrix}$$

where E is the material Young's modulus and v the Poisson's ratio. The variation with respect to the circumferential angle, θ, can be integrated over one radian as

$$c = \int \cos^2 n\theta \, d\theta = \int \sin^2 n\theta \, d\theta = \begin{cases} 1, & n = 0 \\ \frac{1}{2}, & n \neq 0 \end{cases}$$

Note that at this stage the harmonics in the Fourier series uncouple, as other combinations; typically

$$\int \cos n\theta \cos m\theta \, d\theta = \int \cos n\theta \sin m\theta \, d\theta = \int \sin n\theta \sin m\theta \, d\theta = 0$$

all integrating to zero. If the nodal line is the centreline of the shell, then the membrane and bending components also uncouple and the stiffness associated with bending and stretching could be integrated separately. For the element developed here, this assumption is not made. The element stiffness matrix is integrated numerically using Gaussian quadrature[11] with a minimum of four integration points in the ζ direction and two integration points in the η direction.

5.4. ELEMENT LOADS

There are various types of loadings that can be applied to axisymmetric shells. The ones that are of most practical importance are point loads, pressures, self weight and temperature. If the loadings are constant around the circumference of the shell, then they will correspond to the zero harmonic and the majority of loads are of this form. Loadings with an overall sideways resultant can be represented by the first harmonic. More detailed variations of loads around the circumference are expressed in terms of a Fourier series and each harmonic is applied separately.

5.4.1. Pressure Load

The pressure load $p(S, \theta)$ acts on the inner or outer surface of the shell, normal to the surface in the direction of the local displacement $\bar{w}$. From the assumed element interpolation function

$$\begin{aligned} \bar{w} &= [w_1 \; 0 \; 0 \; w_2 \; w_3 \; 0 \; 0 \; w_4 \; w_5 \; 0 \; 0 \; w_5] \mathbf{T}_N \boldsymbol{\rho} \\ &= \mathbf{w}_w \mathbf{T}_N \boldsymbol{\rho} \end{aligned}$$

Using the principle of virtual work, the kinematically equivalent nodal pressure forces are

$$\mathbf{P} = \mathbf{T}_{\mathrm{N}}^{\mathrm{t}} \int_{S} \int_{\theta} \mathbf{w}_{w}^{\mathrm{t}} p(S, \theta) R \, \mathrm{d}\theta \, \mathrm{d}S$$

where R is the radius of the surface on which the pressure is acting. The nth harmonic of the pressure load uncouples from all of the other harmonics in the same manner as the terms in the stiffness matrix do, so that the equivalent loads for the nth harmonic are

$$\mathbf{P} = c\mathbf{T}_{\mathrm{N}}^{\mathrm{t}} \int_{S} \mathbf{w}_{w}^{\mathrm{t}} p_{n}(S) R \, \mathrm{d}S$$

where $p_n(S)$ is the amplitude of the nth Fourier component of the pressure load. This can be defined in terms of the pressure values at the three nodes if the load is to be allowed to vary along the length of the shell:

$$\mathbf{P}_{\mathrm{N}} = \{P_1 \ \ P_2 \ \ P_3\}$$

so that

$$p_n(S) = \mathbf{v}\mathbf{p}_{\mathrm{N}}$$

and the equivalent pressure load is

$$\mathbf{P} = c\mathbf{T}_{\mathrm{N}}^{\mathrm{t}} \int_{-1}^{1} \mathbf{w}_{w}^{\mathrm{t}} \mathbf{v} R \gamma \, \mathrm{d}\zeta \, \mathbf{p}_{\mathrm{N}}$$

If the shell is very thin then the mid-thickness radius can be taken as R. If the pressure is constant along the length of the shell, with magnitude p_{N}, then the equivalent loads are

$$\mathbf{P} = c\mathbf{T}_{\mathrm{N}}^{\mathrm{t}} \int_{-1}^{1} \mathbf{w}_{w}^{\mathrm{t}} R \gamma \, \mathrm{d}\zeta \, p_{\mathrm{N}}$$

Equivalent nodal forces can be found in an exactly similar manner for surface traction forces.

5.4.2. Inertia Loads

In the general case the shell can be subjected to arbitrary accelerations at each nodal freedom so that, for the element

$$\ddot{\mathbf{p}} = \{\ddot{p}_1 \ \ \ddot{p}_2 \ldots \ddot{p}_{12}\}$$

The accelerations at any point are then

$$\ddot{\mathbf{u}} = \begin{bmatrix} \ddot{w} \\ \ddot{u} \\ \ddot{v} \end{bmatrix} = \mathbf{w}\mathbf{T}_{\mathrm{N}}\ddot{\mathbf{p}}$$

and the kinematically equivalent forces are

$$\mathbf{P} = \mathbf{T}_{\mathrm{N}}^{\mathrm{t}} \int_V \mathbf{w}^{\mathrm{t}}\mathbf{w}\rho \, \mathrm{d}V \, \mathbf{T}_{\mathrm{N}}\ddot{\mathbf{p}}$$

where ρ is the density of the shell material. Once again the harmonics all uncouple, so that the inertia forces for the nth harmonic are

$$\mathbf{P} = c\mathbf{T}_{\mathrm{N}}^{\mathrm{t}} \int_{-1}^{1} \int_{-1}^{1} \mathbf{w}^{\mathrm{t}}\mathbf{w}\rho R\gamma\delta \, \mathrm{d}\zeta \, \mathrm{d}\eta \, \mathbf{T}_{\mathrm{N}}\ddot{\mathbf{p}} = \mathbf{M}\ddot{\mathbf{p}}$$

where $\mathbf{M}$ can be identified as the element mass matrix

$$\mathbf{M} = c\mathbf{T}_{\mathrm{N}}^{\mathrm{t}} \int_{-1}^{1} \int_{-1}^{1} \mathbf{w}^{\mathrm{t}}\mathbf{w}\rho R\gamma\delta \, \mathrm{d}\zeta \, \mathrm{d}\eta \, \mathbf{T}_{\mathrm{N}}^{\mathrm{t}}$$

and this can be used to analyse dynamic problems. If the accelerations are constant in a given global direction, then

$$\ddot{\mathbf{p}} = \{a_1 \;\; a_2 \;\; a_3 \;\; 0 \;\; a_1 \;\; a_2 \;\; a_3 \;\; 0 \;\; a_1 \;\; a_2 \;\; a_3 \;\; 0\}$$

where a_1 is the acceleration in the global radial direction, a_2 the acceleration in the axial direction, and a_3 the acceleration in the tangential direction. The acceleration associated with the rotational freedom is zero because the translational accelerations are constant. In this case the equivalent inertia forces are

$$\mathbf{P} = c\mathbf{T}_{\mathrm{N}}^{\mathrm{t}} \int_{-1}^{1} \int_{-1}^{1} \mathbf{w}^{\mathrm{t}}\mathbf{T}\rho R\gamma\delta \, \mathrm{d}\zeta \, \mathrm{d}\eta \, \mathbf{a}$$

where

$$\mathbf{a} = \{a_1 \;\; a_2 \;\; a_3\}$$

and

$$\mathbf{T} = \begin{bmatrix} \sin\phi & -\cos\phi & 0 \\ \cos\phi & \sin\phi & 0 \\ 0 & 0 & 1 \end{bmatrix}$$

where ϕ is the slope of the shell mid-surface at the integration points.

5.4.3. Temperature Loads

The temperature can vary in an arbitrary manner along the length of the shell, but it can only have up to a linear variation through the thickness to be consistent with the thin-shell assumption. This is the same form of variation that was assumed for the shell thickness and, following this, the temperature at any point can be interpolated as

$$H = \tfrac{1}{2}(1-\eta)\mathbf{v}\mathbf{H}_{\mathrm{L_N}} + \tfrac{1}{2}(1+\eta)\mathbf{v}\mathbf{H}_{\mathrm{R_N}}$$

where

$$\mathbf{H}_{\mathrm{L_N}} = \{H_{\mathrm{L_1}} \; H_{\mathrm{L_2}} \; H_{\mathrm{L_3}}\}$$

and

$$\mathbf{H}_{\mathrm{R_N}} = \{H_{\mathrm{R_1}} \; H_{\mathrm{R_2}} \; H_{\mathrm{R_3}}\}$$

are the nodal temperatures on the left-hand and the right-hand surfaces of the shell respectively. The thermal strains at any point on the shell are then

$$\boldsymbol{\varepsilon}_H = \begin{bmatrix} \varepsilon_{SS} \\ \varepsilon_{\theta\theta} \\ \varepsilon_{S\theta} \end{bmatrix}_H = \begin{bmatrix} 1 \\ 1 \\ 0 \end{bmatrix} \alpha H = \boldsymbol{\beta}\alpha H$$

where α is the coefficient of linear thermal expansion

$$\boldsymbol{\varepsilon} = \boldsymbol{\varepsilon}_H + \boldsymbol{\varepsilon}_{\mathrm{E}} = (\boldsymbol{\alpha}_1 + z\boldsymbol{\alpha}_2)\mathbf{T}_{\mathrm{N}}\boldsymbol{\rho}$$

where $\boldsymbol{\varepsilon}_{\mathrm{E}}$ are the elastic strains

$$\boldsymbol{\varepsilon}_{\mathrm{E}} = \boldsymbol{\varepsilon} - \boldsymbol{\varepsilon}_H$$

The stress at any point is then

$$\boldsymbol{\sigma} = \boldsymbol{\kappa}(\boldsymbol{\varepsilon} - \boldsymbol{\varepsilon}_H)$$

which gives the equivalent thermal load as

$$\mathbf{P} = c\mathbf{T}_{\mathrm{N}}^{\mathrm{t}} \int_{-1}^{1}\int_{-1}^{1} (\boldsymbol{\alpha}_1^{\mathrm{t}} + z\boldsymbol{\alpha}_2^{\mathrm{t}})\boldsymbol{\kappa}\boldsymbol{\beta}(\tfrac{1}{2}(1-\eta)\mathbf{v}\mathbf{H}_{\mathrm{L_N}} + \tfrac{1}{2}(1+\eta)\mathbf{v}\mathbf{H}_{\mathrm{R_N}})\alpha R\gamma\delta \, \mathrm{d}\xi \, \mathrm{d}\eta$$

5.5. USING AXISYMMETRIC ELEMENTS

To illustrate the use of axisymmetric thin shell elements and the typical forms of solution, various sample problems are presented in this section. Axisymmetric structures are usually designed to carry loads by membrane action over most of the structure. At any discontinuity in the shell, either in geometry, loading or boundary conditions, there can be local bending effects that arise in the satisfaction of displacement continuity across the structural discontinuity. The peak stresses that arise from such local bending are usually much higher than the membrane stresses. In addition, the bending effects usually die away rapidly from the discontinuity. From classical theory a characteristic length can be defined,[10] where the effects of the discontinuity stresses have become very small. The results reported in the following examples show that the element developed here should have a length, l, which satisfies the condition

$$l < (Rt)^{1/2}/(1-\nu^2)^{1/4}$$

for accurate calculation of the discontinuity stresses. In this t is the local thickness, R is the minimum principal radius of curvature of the shell in the region of the discontinuity and ν is the Poisson ratio of the material. This minimum element length will vary with the element formulation and must be found by experiment. In using axisymmetric thin shell elements there should be a fine mesh in the region of the discontinuity, and a coarser mesh can be used away from here. The transition between the coarse and fine meshes should be gradual for best results; otherwise the transition in the mesh can itself act as a discontinuity. In all of the following examples the same material properties have been used. The values are $E = 210$ GPa, $\nu = 0{\cdot}3$ and $\rho = 7\ \mathrm{Mg/m^3}$.

5.5.1. Axisymmetric Loads (Zero Harmonic)

Loads that are constant around the shell circumference are the most common form of loading for this type of structure, and correspond to the zero harmonic ($n = 0$). In this case, the tangential displacement completely uncouples from the other three displacements and could be omitted from the element formulation. For the zero harmonic the shell can undergo two rigid body motions, a translation parallel to the shell axis and a torsional rotation about the shell axis. This means that at least two supports must be provided for the structure to prevent these rigid body movements. If the position of the supports is arbitrary then they should be placed at the maximum shell radius for the best conditioning of the resulting equations.

Example 1: Cylinder Under Radial Edge Shear

A cylinder of radius 1 m, thickness 0·01 m and length 2 m has a uniform radial shear force of 100 N rad^{-1} applied around the circumference of one end of the cylinder. The other end of the cylinder is taken to be completely fixed. A uniform mesh of 10 elements is used to analyse the structure. A classical thin-shell solution for this problem is given in ref. 12. Figure 5.3 shows a comparison between the finite element solution and exact classical theory for the radial displacement and the rotation along the length of the shell. The agreement is very high everywhere. Figure 5.4 gives a plot of axial and hoop stresses along the length of the shell. These are again in agreement with classical theory (but the comparison is not shown on the figure), with a

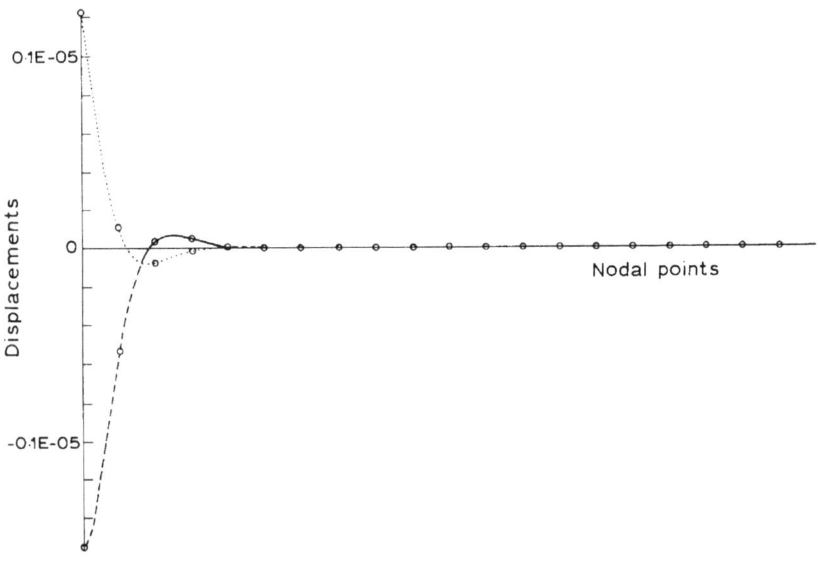

FIG. 5.3. Comparison between the displacement and rotation solutions of the finite element and classical thin-shell theories for a cylinder with an axisymmetric edge shear force. (Displacements in metres.)

		Finite element	Exact
········	Displacement	0·121E – 05	0·122E – 05
– – – – –	Rotation	0·155E – 04	–0·157E – 04
○	Exact		

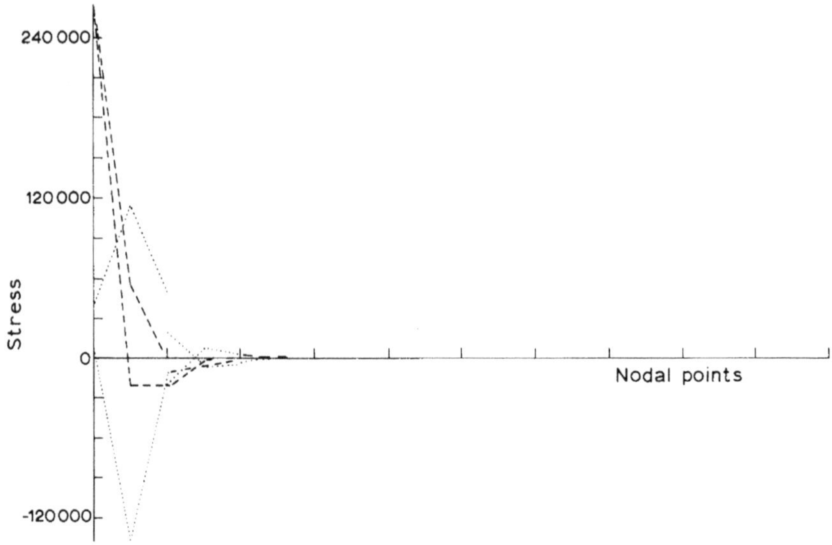

FIG. 5.4. Stress distribution for a cylinder under an axisymmetric radial edge shear force. (Stress in MPa.)

	Minimum	Maximum
--------- Axial	−0·137E + 06	0·114E + 06
- - - - - Hoop	−0·228E + 05	0·266E + 06

difference of 3·5% in the peak stress. For the finite element solution the stresses were calculated directly at the node points and, for plotting purposes, a straight-line fit was used between nodes, which gives an exaggerated 'peaky' representation. This test shows that two elements over the 'die-away' length is sufficient to achieve a good accuracy in the results. A finer mesh (40 elements) gives results that are indistinguishable from the exact solution, with no significant discontinuity in the stresses across elements.

Example 2: Cylinder/Sphere Under Internal Pressure

A sphere of radius 1 m is connected to a cylinder of length 1·5 m. The complete shell has uniform thickness of 0·025 m and is loaded by a uniform internal pressure of 1 MPa. The lower end of the cylinder is

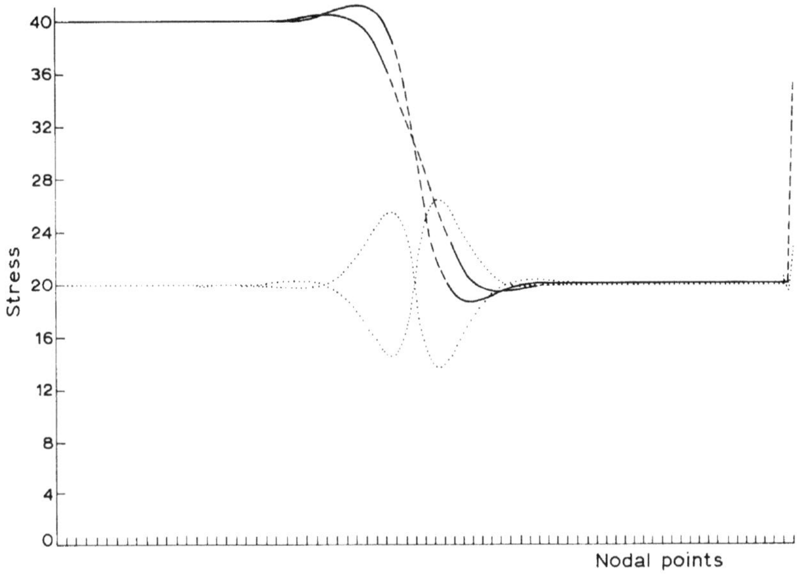

FIG. 5.5. Stress distribution for a cylinder/sphere shell with an internal pressure of 1 MPa. (Stress in MPa.)

	Minimum	*Maximum*
--------- Axial	13·7	26·3
– – – – – Hoop	18·6	41·2

fixed against vertical and tangential displacements to prevent rigid body movements. A mesh of 76 elements is used, with 36 elements along the cylinder and 40 elements around the sphere. The actual nodal positions along the surface of the shell are shown on the horizontal axis of Fig. 5.5. The distribution of axial and hoop stresses around the surface of the shell is also shown in the figure. It will be seen that, away from the discontinuity, the stresses are equal in the sphere, and the hoop stress is twice the axial stress in the cylinder. The discontinuity effects at the junction exhibit the same 'die-away' length in both the cylinder and the sphere, so that the local cylindrical radius can be used to estimate the element density in any shape of shell. There is an error in the stress calculation where the sphere touches the axis of symmetry, because the radius is zero here, making the hoop strain calculation ill-conditioned. The error only occurs at the node on the centreline and the stress there can be ignored. Alternatively, for the element that touches the centreline, the stresses can be calculated at the Gaussian integration points and extrapolated to the nodes. The results shown here are for a fine mesh and agree exactly with classical thin-shell theory. In practice, a coarser mesh would be used, with local refinement in the region of the junction.

5.5.2. Non-axisymmetric Loads—First Harmonic

The axisymmetric element can also be loaded by forces that vary around the circumference of the shell. For the first harmonic (with $n = 1$) these forces will have an overall resultant. The radial and tangential forces have a sideways shear resultant (Fig. 5.6) and the axial and moment forces have an overall moment resultant (Fig. 5.7). Resolving and integrating the forces in Fig. 5.6 gives the two shear resultants as:

$$S_A = \pi(P_{11} - P_{31})$$
$$S_B = \pi(P_{12} + P_{32})$$

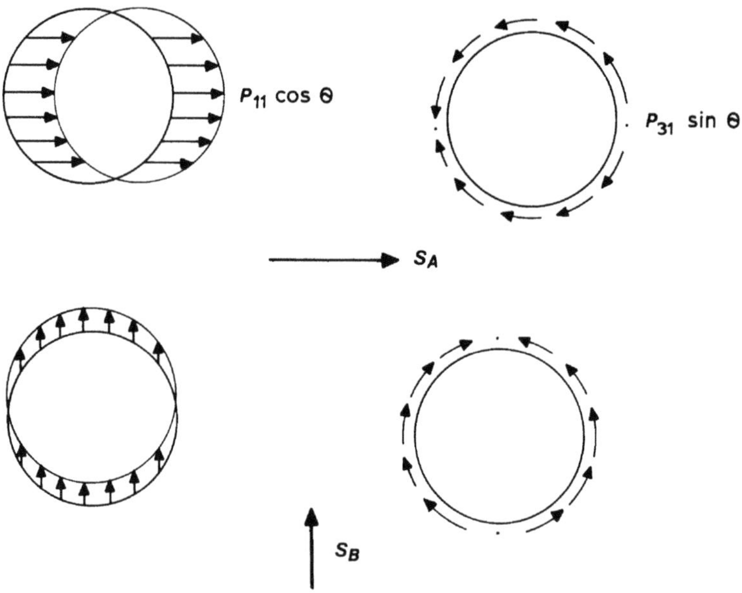

FIG. 5.6. Shear resultants for harmonic number 1.

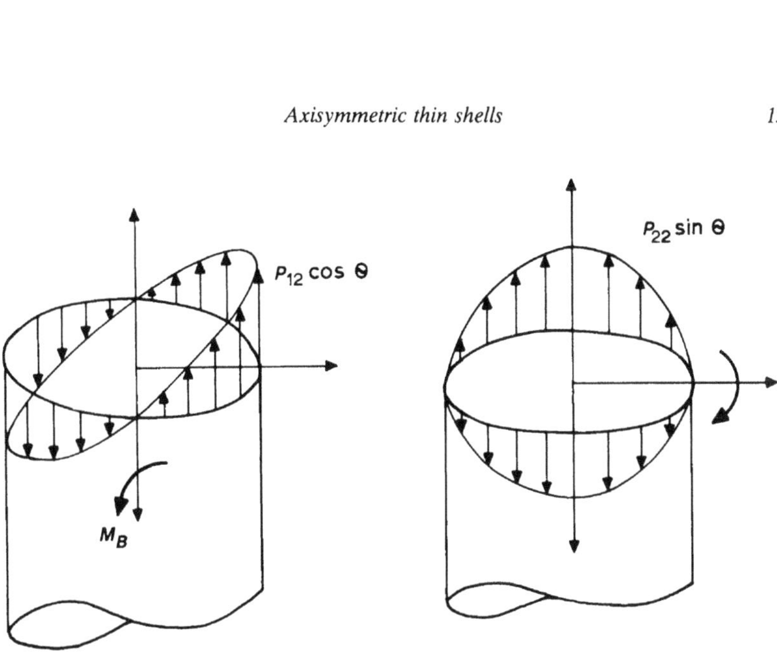

FIG. 5.7. Moment resultants for harmonic number 1.

For a one-radian slice of the shell a horizontal force, S_A, can be applied in an infinite number of ways, typically:

$$S_A = \pi P_{11}$$

$$S_A = -\pi P_{31}$$

$$S_A = \frac{\pi}{2}(P_{11} - P_{31})$$

The shell can be supported against a sideways rigid body motion by suppressing a radial displacement, suppressing a tangential displacement, and both together. From Fig. 5.7 the forces can be integrated to give overall moment resultants:

$$M_A = \pi r P_{22}$$

$$M_B = \pi r P_{21}$$

In the general case the first harmonic has four possible rigid body movements, translations corresponding to S_A and S_B, and rotations corresponding to M_A and M_B. Very often, the element is only coded with half of the full set of displacements, typically $\rho_{11} \cos n\theta$, $\rho_{21} \cos n\theta$, $\rho_{31} \sin n\theta$ and $\rho_{41} \cos n\theta$, since the other set are identical to these but rotated by 90 degrees. The two sets never couple for linear problems. If this is the case there are only two rigid body

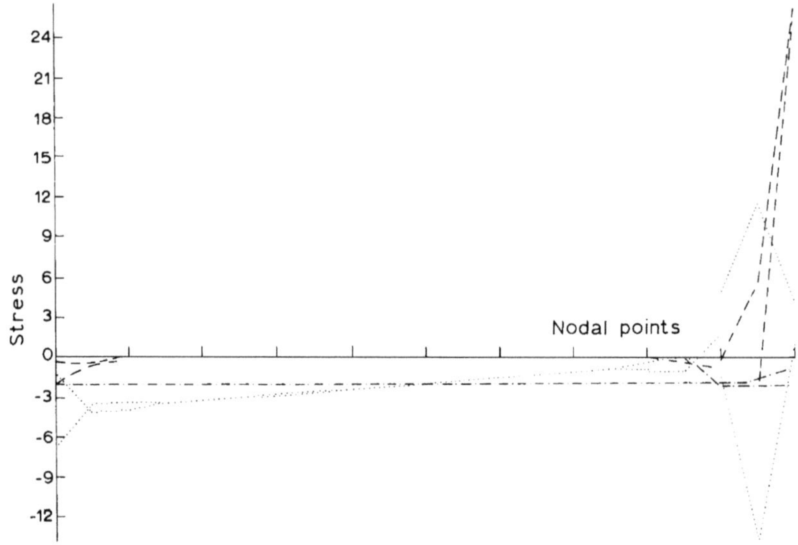

FIG. 5.8. Stress distribution for a cylinder with an edge shear resultant represented by radial and tangential forces with radial, axial and tangential end support. (Stress in MPa.)

	Minimum	*Maximum*
--------- Axial	−0·139E + 07	0·112E + 07
– – – – – Hoop	−0·227E + 06	0·265E + 07
–·–·–·– Shear	−0·236E + 06	−0·904E + 05

movements for $n = 1$, one sideways translation and one overall rotation.

Example 3: Cylinder as a Cantilever Beam With an Edge Shear

This example can be used to investigate the effects of different possible representations of the loads and the effects of different boundary conditions. The same cylinder geometry as Example 1 is used here, and Figs. 5.8–5.10 show the stress distribution for various combinations of loads and boundary conditions. Figure 5.8 is for a shear force represented by

$$S_A = \frac{\pi}{2}(P_{11} - P_{31})$$

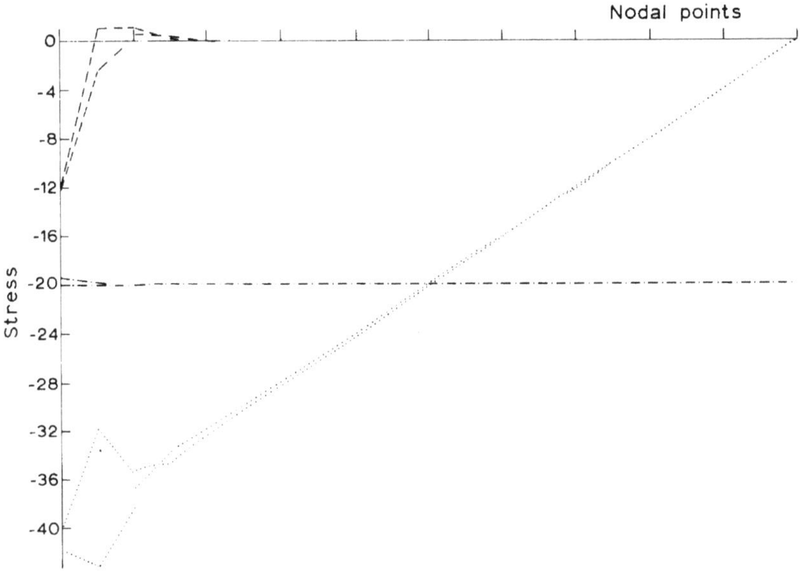

FIG. 5.9. Stress distribution for a cylinder with an edge shear resultant represented by a tangential force with radial, axial and tangential end support. (Stress in MPa.)

	Minimum	Maximum
--------- Axial	−0·432E + 06	−0·858E + 01
– – – – – Hoop	−0·125E + 06	0·107E + 05
–·–·–·– Shear	−0·202E + 06	−0·195E + 06

where $P_{11} = P_{31} = 1000\,\mathrm{N\,rad^{-1}}$. All of the freedoms at the built-in end are fixed. It can be seen that there are localised regions of high stress at both ends of the shell. The 'die-away' length for these is of the same order as that for the zero harmonic of Fig. 5.3. Over the central length of the shell the shear stress is constant, the axial stress is varying linearly and the hoop stress is zero. In this portion of the shell the results correspond to Engineers Theory of Bending (ETB). Figure 5.7 shows the same shell, with the same support conditions but with the load applied as $S_A = -\pi P_{31}$ and $P_{31} = -1000\,\mathrm{N\,rad^{-1}}$. In this case it will be seen that there is no local stress concentration at the loaded end. Figure 5.10 has the same model for the applied load but in this case only the axial and tangential displacements are fixed at the supported

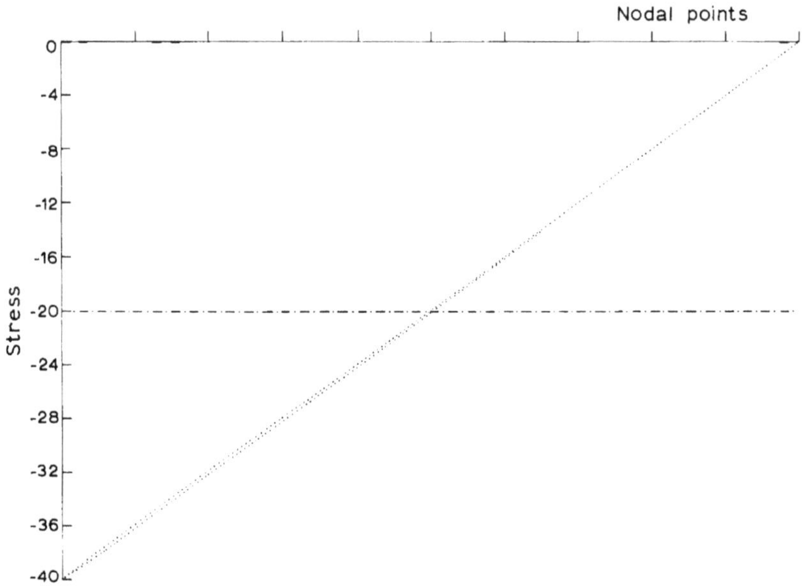

FIG. 5.10. Stress distribution for a cylinder with an edge shear resultant represented by a tangential force with axial and tangential end support. (Stress in MPa.)

	Minimum	Maximum
--------- Axial	−0·400E + 06	−0·858E + 01
- - - - - Hoop	−0·172E + 04	0·430E + 03
–·–·–·– Shear	−0·200E + 06	−0·200E + 06

end. Here there are no local stress concentrations at either end, and the solution agrees exactly with ETB for both stresses and displacements provided the shear deflection is included. These results were obtained with a coarse mesh of 10 equal length elements. The difference in the peak stresses was not more than 6% when compared to a fine mesh of 40 uniform elements.

Example 4: Cylinder Under Self Weight

The cantilever cylinder of the previous example can also be analysed for self-weight loads. Using the support conditions of no axial or tangential displacements at the supported end, and applying the load in the form $\rho a_1 = -\rho a_3 = 100$ then the solution shown in Fig. 5.11 is

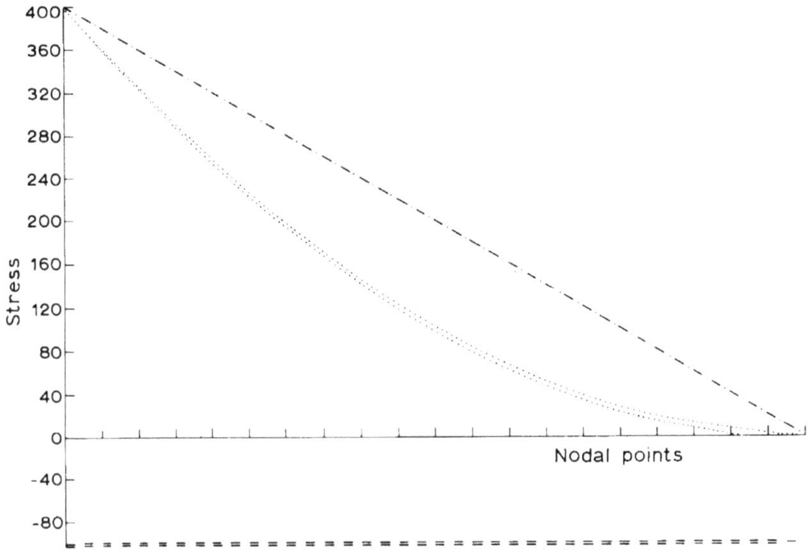

FIG. 5.11. Stress distribution for a cantilever cylinder under self weight (axial and tangential end support). (Stress in MPa.)

	Minimum	Maximum
--------- Axial	−0·899	400·0
– – – – – Hoop	−0·102	98·5
–·–·–·– Shear	−0·499	400·0

obtained. The linear variation of shear stress and the parabolic variation of axial stress corresponds exactly to the ETB solution. It is interesting to note that this solution gives a constant hoop stress along the cylinder (which gets relatively smaller as the aspect ratio of the cylinder is increased). The same overall resultant force can be obtained by applying an acceleration load of $\rho a_3 = -200$. This gives exactly the same distribution of axial and shear stress but now the hoop stress is zero.

Example 5: Pseudo-static Seismic Load on a Cylinder/Sphere Geometry

The geometry of Example 2 is loaded by a sideways acceleration. The base of the cylinder was taken as being fixed against vertical and

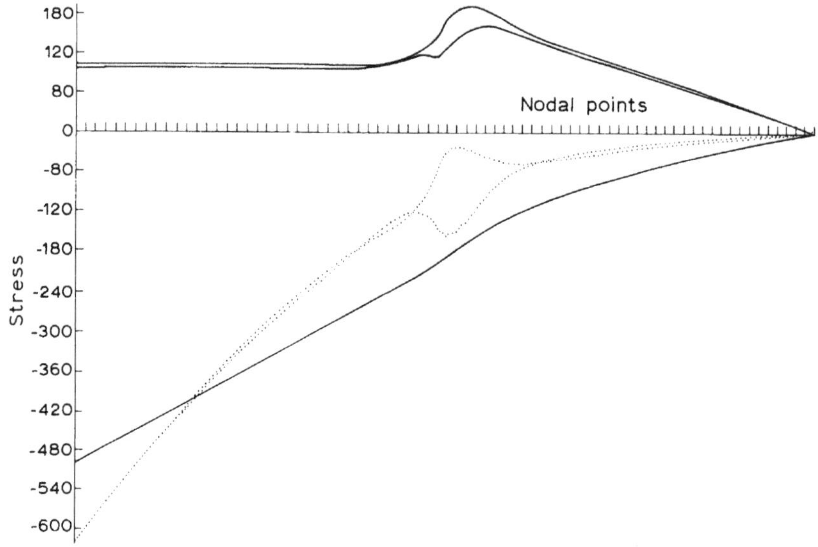

FIG. 5.12. Stress distribution in a cylinder/sphere shell under a uniform sideways acceleration. (Stress in MPa.)

		Minimum	Maximum
....	Axial	−625·0	1·40
———	Hoop	−3·87	190·0
- - - - -	Shear	−501·0	1·35

tangential movements. The resulting stress distribution is shown in Fig. 5.12. The results in the cylindrical portion of the shell are similar to those obtained in the previous example. There is a small discontinuity effect at the junction between the cylinder and the sphere. The stresses in the sphere are zero on the centreline, as they should be from the antisymmetric nature of the loading. In this case there is no spurious stress on the centreline as there was in Example 2, even though the same form of numerical ill-conditioning still exists, but it does appear with an increasingly coarse mesh.

5.5.3. Non-axisymmetric Loads—Higher Harmonics

If the harmonic number, n, is greater than 1, then all of the loads applied to the shell are self-equilibrating and have no overall resultant.

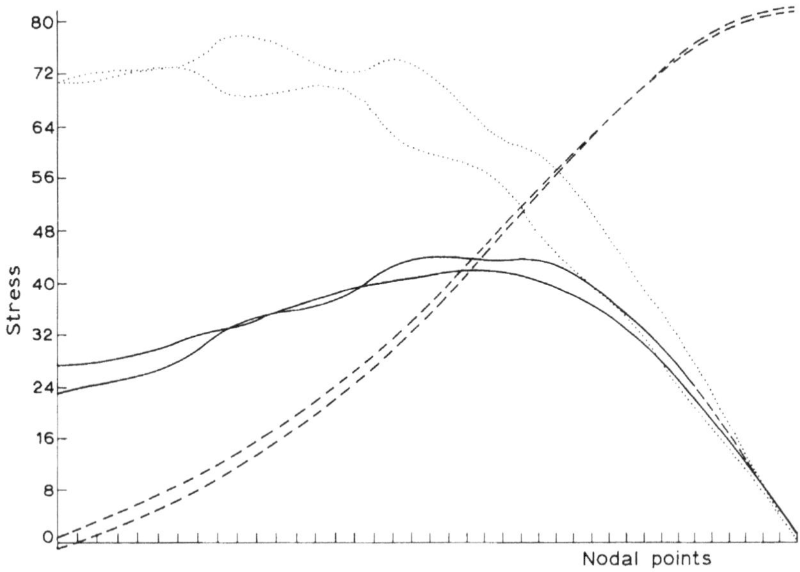

FIG. 5.13. Stress distribution in a hyperbolic shell with a local axial end tension of cos 2θ. (Stress in MPa.)

	Minimum	Maximum
---------- Axial	15·8	77·8
———— Hoop	1·03	43·8
- - - - - Shear	−0·959	82·0

This means that there are no possible rigid body motions for $n > 1$ and no supports are required. Most shells are very much more flexible for $n = 2$ than they are for $n = 0$ or 1 and, consequently, have larger deflections for a given load. As the harmonic number increases they become progressively stiffer again. The harmonic number for which the stiffness is an absolute minimum depends upon the geometry of the shell. If only the deflections are required then relatively few harmonics of a general loading need be analysed. If stresses are required then convergence can be much slower and sometimes a considerable number of harmonics are required. The criteria here is the rate of change of stress in the circumferential direction for the real problem. If this is rapid then many harmonics are required.

Example 6: Hyperbolic Shell Under a Non-axisymmetric Edge Load
A shell with negative Gaussian curvature has a cross-section defined by the hyperbola $x^2 - y^2 = 1$. Its thickness is 0·01 m and it has a length of 1 m from the waist circle, and has been analysed subject to an horizontal load of $0{\cdot}5774 \cos 2\theta$ and a vertical load of $0{\cdot}8165 \cos 2\theta$ at the end away from the waist. The shell is supported against tangential displacement only at the waist. For classical thin-shell theory[13] all three stresses are pure membrane stresses for this loading. The finite element solution is shown in Fig. 5.13, where it will be seen that the stresses are predominantly membrane but not completely so. There is a difference of 0·4% between the maximum stress from the finite element and the classical solutions. Figure 5.14 shows the finite

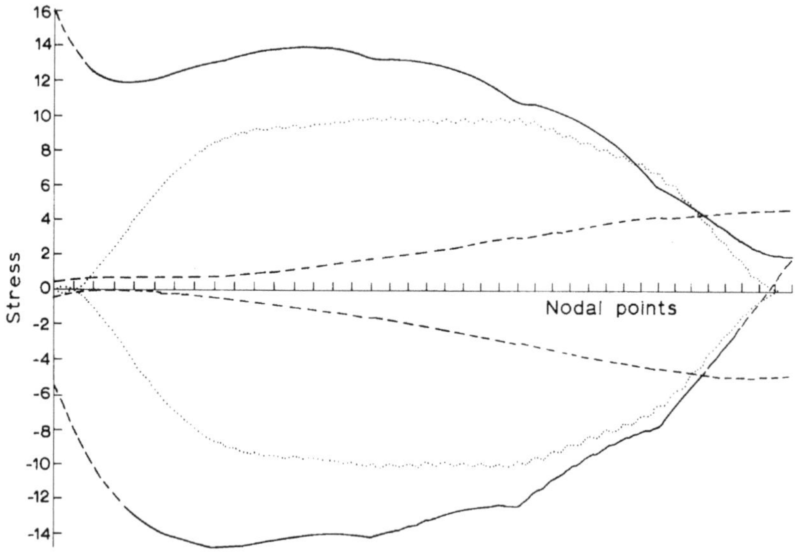

FIG. 5.14. Stress distribution in a hyperbolic shell with a radial edge force of $0{\cdot}5774 \cos 2\theta$, tangential waist displacement fixed. (Stress in MPa.)

		Minimum	Maximum
----------	Axial	−0·101E + 05	0·998E + 04
————	Hoop	−0·148E + 05	0·181E + 05
– – – – –	Shear	−0·498E + 04	0·474E + 04

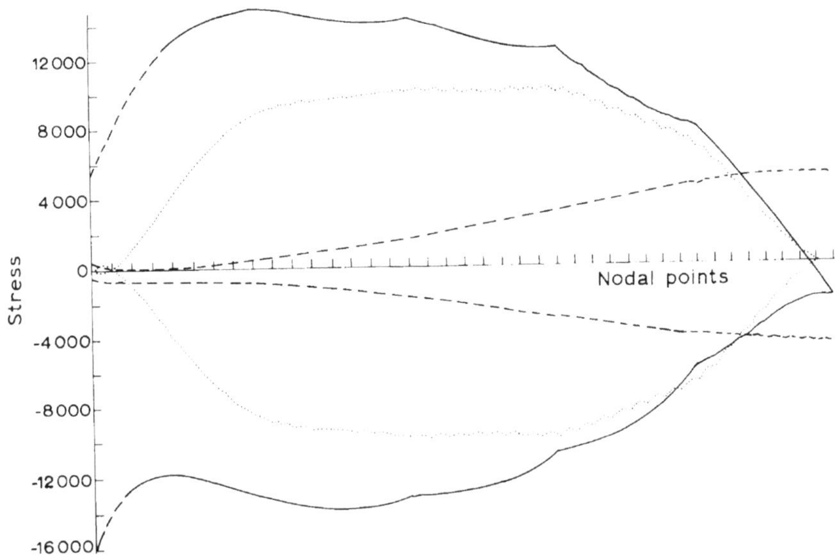

FIG. 5.15. Stress distribution in a hyperbolic shell with a vertical edge force of 0·8165 cos 2θ, tangential waist displacement fixed. (Stress in MPa.)

	Minimum	Maximum
--------- Axial	−9 890	10 200
——— Hoop	−16 100	14 900
- - - - - Shear	−4 660	5 060

element solution for the horizontal component of the load and Fig. 5.15 shows the solution for the vertical component. It will be seen that these are two orders of magnitude greater than the combined results of Fig. 5.13. The fact that the stresses are not pure membrane stresses in the finite element solution is due to the modelling approximations, rounding error, and the cancellation of the bending effects not being exact. One point to notice from Figs 5.14 and 5.15 is that for a shell with negative Gaussian curvature the stresses do not necessarily die away along the length of the shell. A slight change to the support conditions, where now the vertical displacement is also fixed at the waist results in the stress distributions shown in Figs 5.16 and 5.17 for the individual components of the force. It will be seen that a relatively

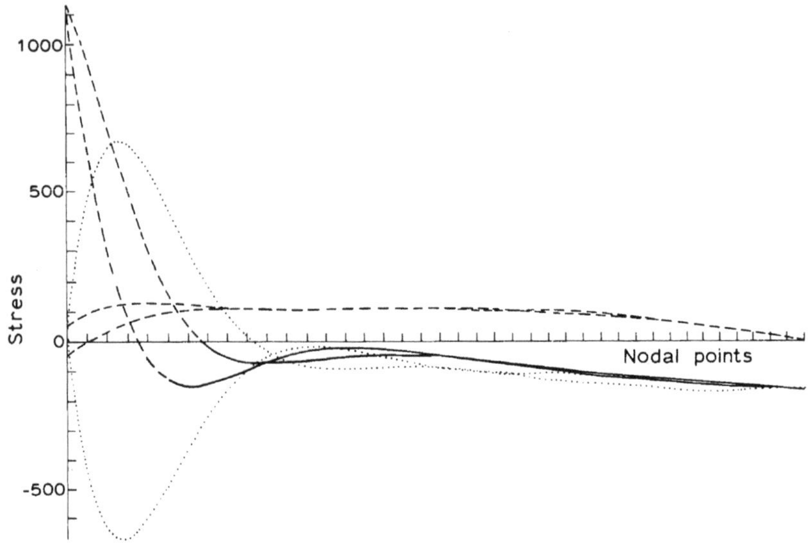

FIG. 5.16. Stress distribution in a hyperbolic shell with a radial edge force of $0{\cdot}5774\cos 2\theta$, vertical and tangential waist displacements fixed. (Stress in MPa.)

		Minimum	*Maximum*
----------	Axial	−673·0	664
———	Hoop	−178·0	1 140
- - - - -	Shear	−48·9	126

small change in the support conditions has resulted in major changes in the stress distribution and now the stresses do have a 'die away' length associated with them. Interestingly, the combined loading with the modified boundary conditions results in almost exactly the same stress distribution as Fig. 5.13.

5.6. DYNAMICS

The axisymmetric element can be used directly for a dynamic analysis. The equations of motion are

$$\mathbf{M}\ddot{\mathbf{r}} + \mathbf{K}\mathbf{r} = \mathbf{R}(t)$$

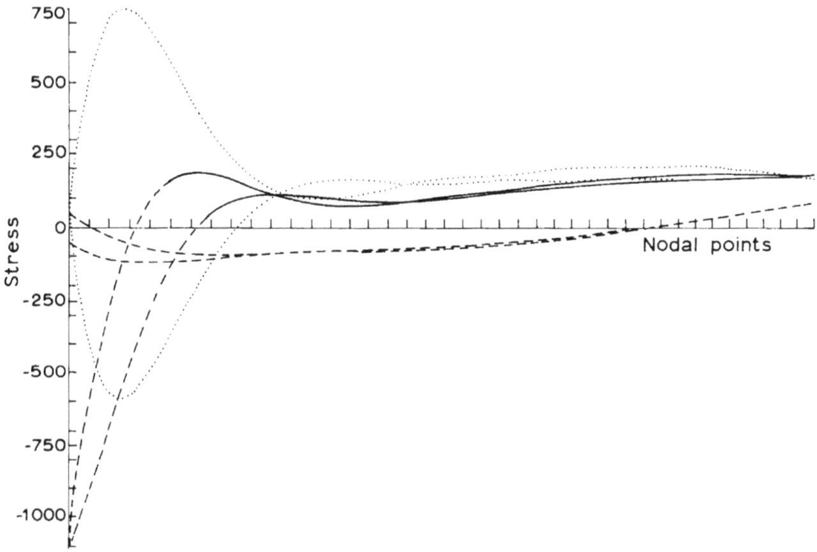

FIG. 5.17. Stress distribution in a hyperbolic shell with a vertical edge force of $0{\cdot}8165\cos 2\theta$, vertical and tangential waist displacements fixed. (Stress in MPa.)

		Minimum	*Maximum*
....	Axial	−583	746·0
———	Hoop	−1 110	188·0
– – – – –	Shear	−122	81·7

The assembled stiffness matrix is exactly as for the static case and the mass matrix can be assembled from the element mass matrices that were derived in Section 5.4.2. One major advantage of the axisymmetric element is that the circumferential harmonics uncouple for vibrations, so that the natural frequencies and mode shapes can be found independently for each harmonic. A dynamic calculation is then conducted as a series of small models (one for each circumferential harmonic) rather than as a single large problem that includes all of the harmonics. This makes the calculation of normal modes and frequencies much cheaper than for a single large analysis. It is usually found that the lowest frequency of vibration is associated with a circumferential harmonic number in the range of 2 to 6 depending upon the actual geometry.

5.7. NON-LINEAR ANALYSIS

The axisymmetric element can be used for plasticity and large-deflection analysis provided the assumption that the shell remains substantially circular holds. The main efficiency of the axisymmetric elements, that the harmonics are all uncoupled, is lost. Typically, for an elastic–plastic analysis the stiffness matrix will be of the form

$$\mathbf{k} = \mathbf{T}_{\mathrm{N}}^{\mathrm{t}} \int_V \boldsymbol{\alpha}^{\mathrm{t}} \boldsymbol{\kappa}(\sigma) \boldsymbol{\alpha} \, \mathrm{d}V \, \mathbf{T}_{\mathrm{N}}$$

where $\boldsymbol{\kappa}(\sigma)$, the material stiffness, is a function of stress. Expanding the forces, the displacements and now also the stresses in terms of a Fourier series results in integrals of the form

$$\int \cos l\theta \cos m\theta \cos n\theta \, \mathrm{d}\theta$$

and there are many combinations of l, m and n for which the integration is non-zero.[5] The harmonics couple together and they cannot be solved as separate problems. The size of the matrices becomes proportional to the number of harmonics used in the analysis and, as the problem becomes more non-linear, the greater becomes the coupling. The equations should be assembled so that, for a given node, all of the freedoms for all of the harmonics are grouped together to minimise the bandwidth of the equations.

A similar coupling between the modes arises if large deflections are to be allowed. This is especially important if a buckling analysis is being considered. A shell under axisymmetric loadings can buckle into a non-axisymmetric mode shape. It is necessary then to investigate a range of harmonics to find the one which gives the absolute lowest buckling load. It is also worth noting that the buckling of many axisymmetric shells is imperfection sensitive and the observed buckling load can be considerably less than the calculated value. This sensitivity is pointed to by Example 6, where it was observed that a slight change in the support conditions could lead to very large changes in the stress distribution.

5.8. APPROXIMATELY AXISYMMETRIC GEOMETRIES

An axisymmetric model can also be used to analyse structures that are only approximately axisymmetric. If the modulus of elasticity varies around the circumference of the shell, then this can be expanded in terms of a Fourier series in θ. The integration required to form the element stiffness matrix will then couple all of the harmonics together, exactly as for the non-linear analysis, so that the response to all of the harmonics must be found simultaneously. Other situations can arise where the structure is essentially axisymmetric but with some circumferential variation. Typically, this can involve longitudinal stiffeners in shells or bolt holes in flanges. If there are a large number of these around the circumference then their result can be smeared (usually on an area basis), so that an equivalent thickness or equivalent Young's modulus is used in a full axisymmetric analysis. In other cases the change in stiffness around the circumference can be expressed in terms of a Fourier series in θ and the coupled problem can be solved.[14] Where possible, the axisymmetric smearing idealisation should be used, since this allows each harmonic to be solved separately.

5.9. CONCLUSIONS

An axisymmetric thin-shell finite element has been developed and applied to a series of problems. These have shown that very good results can be obtained. All the time that the circumferential harmonics remain uncoupled, this form of solution is economical. The examples did not include any branched shells, but the formulation of the element is such that these can be solved directly without any modifications being required. Examples of branched shells were not included because no simple solutions exist for the purposes of illustrating the behaviour of the element. For non-linear problems and geometries that are almost axisymmetric, the element can still be used but now the harmonics tend to couple together. In this case it is debatable whether it is better to use the axisymmetric element or to solve the problem using a general shell element.

REFERENCES

1. Zienkiewicz, O. C., *The Finite Element Method,* 3rd edn, McGraw-Hill, New York, 1977.
2. Grafton, P. E. and Strome, D. R., Analysis of axisymmetric shells by the direct stiffness method, *AIAA J.*, **1,** 1963, 2342–7.
3. Klein, S., A study of the matrix displacement method as applied to shells

of revolution, *Proc. Conf. on Matrix Methods in Structural Mechanics,* Air Force Inst. Tech., Wright Patterson Air Force Base, Ohio, Oct. 1965.

4. Chan, A. S. L. and Firmin, A., The analysis of cooling towers by the matrix finite element method, *Aeronaut. J.,* **74,** 1970, 826–35.
5. Chan, A. S. L. and Trbojevic, V. M., Thin shell finite element by the mixed formulation, *Comp. Meth. Appl. Mech. Engng:* Part 1, **9,** 1976, 337–67; Parts 2 and 3, **10,** 1977, 75–103.
6. Klein, S. and Sylvester, R. J., The linear elastic dynamic analysis of shells of revolution by the matrix displacement method, *Proc. Conf. on Matrix Methods in Structural Mechanics,* Air Force Inst. Tech., Wright Patterson Air Force Base, Ohio, Oct. 1965.
7. Bushnell, D., Stress, stability and vibration of complex branched shells of revolution, *Comput. Struct.,* **4,** 1974, 399–435.
8. Bushnell, D., BOSOR5—program for buckling of elastic–plastic complex shells of revolution including large deflection and creep, *Comput. Struct.,* **6,** 1976, 221–39.
9. Cantin, G., Strain displacement relationships for cylindrical shells, *AIAA J.,* **9,** 1968, 1787.
10. Dym, C. L., *Introduction to the Theory of Shells,* Pergamon Press, Oxford, 1974.
11. Shied, F., *Numerical Analysis,* Schaum Series, McGraw-Hill, New York, 1968.
12. Timoshenko, S. and Woinowsky-Krieger, *S., Theory of Plates and Shells,* McGraw-Hill, New York, 1959.
13. Novoshilov, V. V., *The Theory of Thin Shells,* Noordhoff, 1959.
14. Bushnell, D., Stress, buckling and vibration of prismatic shells, *AIAA J.,* **9,** 1971, 2004–13.

Chapter 6

Finite Element Analysis and the Design of Thin-walled Ship Structures

DAVID ANDREWS

*Sea System Controllerate, Ministry of Defence, Bath, UK**

6.1. INTRODUCTION

This chapter considers finite element analysis in the design of thin-walled ship structures. The type of ship which is primarily addressed is the conventional monohull; however, as Caldwell[1] recently illustrated, this is by no means the only form of marine vehicle. Even disregarding the steel-based thin-walled structures in the offshore industry dealt with in Chapter 7, this leaves a plethora of vehicle types, ranging from surface-effect vessels to large military submarines, of which the monohulled displacement craft happens to be the most common and diverse. In just about every category of marine vehicle, the finite element method (FEM) has been adopted for their structural design, with varying degrees of sophistication and success. This chapter indicates, through examples, the range of applications of FEM to thin-walled structures in 'ships' in the broadest sense. The chapter shows that there is a commonality in the application, regardless of ship type and the FEM, as the most common and powerful structural analysis tool, has to be seen within the context of the overall ship design task.

The chapter commences with the ship structural design problem, which has changed in complexity and sophistication over the past 20 to 30 years. The place of the FEM in this development has been

* Formerly at Department of Mechanical Engineering, University College London, UK.

significant and its impact, both in the understanding of ship structures and still more recently as a readily usable *design* technique, is profound. Secondly, the nature of the many loads on ship structures is considered since, for a vessel on the water–air interface, this constitutes the major structural uncertainty. Thus, any technique modelling structural behaviour must recognise the relevant problem in defining structural loads on ships. Loading is the essential first step in structural design and has to be combined with ship strength or structural capability. Through this combination, interaction or response, various modes of structural failure can occur and, clearly, any structural analysis technique must be capable of predicting those failure modes of most significance in the given structural application under investigation. The manner in which structural adequacy is assessed is discussed, as this is not just the measure of the design intent, in the inevitable trade-offs for economic efficiency, but it is also the measure of merit used in approaches to structural optimisation, in which the FEM provides an essential lynch-pin.

Having considered the essentials of ship structural design with regard to the impact the FEM has made to its practice, it is then possible to consider more directly the application of the FEM to ship structures. This is done by first outlining various initial approaches from which many of the current limitations and benefits were first identified. In considering recent and current applications of the FEM to ship structures, a distinction is drawn between FEM as a tool of analysis, both of the structural response of the whole ship structure and specific vital components of the structure, and the FEM directly applied to the *design* of the ship's structure. This is normally undertaken with the finite element system as an integral element of a powerful Computer Aided Ship Design (CASD) system; furthermore, this may also then be tied into a CADCAM system.

The chapter concludes with consideration of the current and future development in the application of the FEM to ship structures, together with an outline of the ground rules in the use of FEM in the design of ship structures.

It needs to be emphasised that the chapter is deliberately restricted to the overall structural behaviour of ships or marine vehicles. Thus, it excludes consideration of the application of the FEM to the analysis of components and equipment on ships, some of which are sizeable (e.g. 250 tonne gearboxes) or very complex (such as pressure vessels). This limitation is justified both in that the application of finite element analysis to such components on ships differs little from the application in other similar fields of engineering, and also that the FEM used are not usually encompassed within the scope of the thin-walled structure approach that is the subject of this book.

6.2. THE SHIP STRUCTURAL DESIGN PROBLEM

In order to appreciate the way in which the FEM's have altered, and will continue to change, the ship designer's perception of ship structural design, it is necessary to consider the problem the ship designer has to tackle in designing a new structure for a new ship. Thus, it is necessary to see the structural design as part of the total ship design problem and while traditionally it has been possible for the structural design, by virtue of both its lack of sophistication and the provision of precise rules[2] from the Classification Societies, to be isolated in the design evolution, this has definitely changed in recent years. There are several parallel reasons why this occurred:

(a) new 'ship' types and materials rendered 'rule of thumb' and Society Rules inapplicable;
(b) the increased demand for 'cost-effective' total designs meant the cost of ship structure as proportion of ship cost had to be reduced;
(c) increased knowledge of structural loading, structural response and structural failure;
(d) wide-scale use of comprehensive CASD systems, incorporating structural analysis, performed largely by means of FEM.

This has meant that the changing view of ship structural design, leading to what is commonly called a 'Rational' approach to ship structural design, has been more related to the overall aim of a given ship design than had previously been the case. As will be seen the role that finite element techniques play in the Rational approach is highly significant at several component stages in the approach. However, its importance must not be over-emphasised, as extensive analysis without comparable in-depth knowledge of loading, together with a believable criterion of behaviour, means that the investment in the analysis will be largely wasted. The final aspect to be considered in ship structural design is the nature of structural synthesis, and this means that the designer has to adopt a clear model of the structural behaviour of the ship.

Ship structural design is one of the primary considerations in the design of a ship, which have been conveniently categorised as S^5, i.e. Speed, Sea-keeping, Stability, Structure and Style. Of these categories, the first two are aspects of ship performance related to the customer's requirements, the second two are essential to the safety of the vessel, which the customer largely takes for granted, and the final category relates to a host of less quantifiable aspects such as operational performance, protection, habitability, margin policy and the general style of a given design. These latter aspects are highly dependent on ship type and role or market.

As a ship design evolves, all the various aspects covered by the categories are gradually resolved, so the feasibility and then the reliability of the design emerges. However, the resolution is not within separate spheres; most of the aspects interact with each other and a decision to change the approach for one aspect, say top speed, can influence the solution of many other features in the final design. Considering a structural example, the choice of style can be significant, thus a robust more rugged solution requiring less maintenance will mean heavier steelwork. Given that steelwork is significant the overall weight (e.g. 50% of the lightweight and 35% of the displacement of a frigate, or 12% of the weight of a very large tanker), this will lead to the growth in displacement due to spiralling nature of knock-on effects, e.g. bigger engines to maintain speed, more fuel, bigger ship, heavier structures, bigger engine, etc. This spiral not only exhibits the interactive nature of the ship design process but also the pressure on the structural designer to reduce the impact of the steelwork on the overall designs.

Having said that the various aspects of ship design interact, the manner in which they do so is far from clear in a generalised sense and so the design process proceeds in a somewhat cautious way. The process is one of integrating not just the mass of equipments and systems that go to make the ship but also the numerous interrelated features necessary to meet the aspects of the design covered by the S^5 topics already mentioned. Thus the structural design process has to be seen as an interactive, interdependent component of the ship design task and, as such, many of the constraints on the structural design are non-structural in origin.

Consider the basic configuration of the overall structure. This arises from several constraints, e.g.:

(a) environmental such as watertightness, seaworthiness;
(b) operational, such as cargo handling (e.g. container hatches), aircraft operations (from landing platforms to the ultimate of an aircraft carrier), hull form for high speed;
(c) legislations constraints laid down by national bodies, e.g. Department of Trade (DOT) or international bodies, e.g. International Maritime Organisation (IMO), and Safety of Life at Sea (SOLAS), laying down stability standards which constrain the ship's beam.

Those types of constraints result in the conventional monohull having a structure characterised by:

(1) An external form that is elongated, watertight, with its longest dimension in the direction of motion, symmetrical about its longitudinal centreline plane, at least below the waterline, superstructure and upper works discontinuous and a complex underwater form.
(2) An internal structure that is hollow, and subdivided by decks and bulkheads with many discontinuities.

This is far from ideal in a structural sense, and so the structural designer has to appreciate how he is constrained by the wider aims in the design. There are also the more familiar constraints on any structural designer, namely those of the manufacturing process and material limitations. In the case of most ships these are those of a large-scale steel fabricating process using semi-automatic forming and welding techniques. Thus a completed structural design is the outcome of compromise rather than a single-minded solution to structural excellence.

The developments listed at the beginning of this section which have led to the structural design of ships becoming more integrated into the general ship design process have been the same developments that have led to ship structural design becoming more scientific in the selection of materials and scantlings.[3] For example the post-war innovations in ship types include giant tankers, articulated ships, high-speed container ships, offshore support vessels and unconventional vehicles (e.g. hovercraft, hydrofoils, SWATH—small water plane twin-hulled). In many instances previous, empirically based, design methods are wholly inadequate. In parallel, new materials such as high-tensile steel, reinforced plastics and reinforced concrete have demanded better understanding. Also, knowledge of the ocean environment and the ship's structural response has improved so the insights from detailed structural analysis of ships by FEM has grown alongside the 'Rational' approach to ship structural design.

The seminal event in the adoption of a Rational approach can be taken as the convening of the first International Ship Structures Conference (ISSC) in 1961. Over the years this body has encouraged the classification societies to move from the specific tabulated structural scantlings selected by type and size of ship to an approach outlined in Fig. 6.1.[4] This approach considers the process of ship structural design in logical steps which are further explained below:

(1) *Formulation of requirements.* This is intended to be a broad statement of the necessary capabilities required of the structure, and must relate not just to the main loadings but also the basis upon which the structure will be evaluated.
(2) *Selection of topology and form.* As already discussed, this is largely determined by non-structural considerations, but there are exceptions to this and choice of material can have a profound impact.
(3) *Choice of scantlings.* The final choice will emerge from the total process, but it will be constrained by the available plate and section sizes available in the material chosen (assuming a steel structure for the present) and the fabrication facility selected.
(4) *Mathematical description of behaviour.* The production of a mathematical model of the ship's structure will be required at different stages of the design, and the model type selected will be dependent on what is required at that stage of the design, the relevance of the various components of the structure to the precise requirement and the actual structure being described.
(5) *Criteria for acceptability.* The designer has to have a clear basis on which he will judge the response of the model to the perceived loads with stated acceptable limits of response.
(6) *Analysis of structure.* There are three principal requirements to obtain an adequate analysis; representative loading on the structure, calculation of the structural response to the loading, and determination of the limiting capability or strength of the structure.
(7) *Criteria of acceptability.* On the basis of the likely modes of failure of the structure to the expected loading, the structure has to be designed with clear margins against failure.

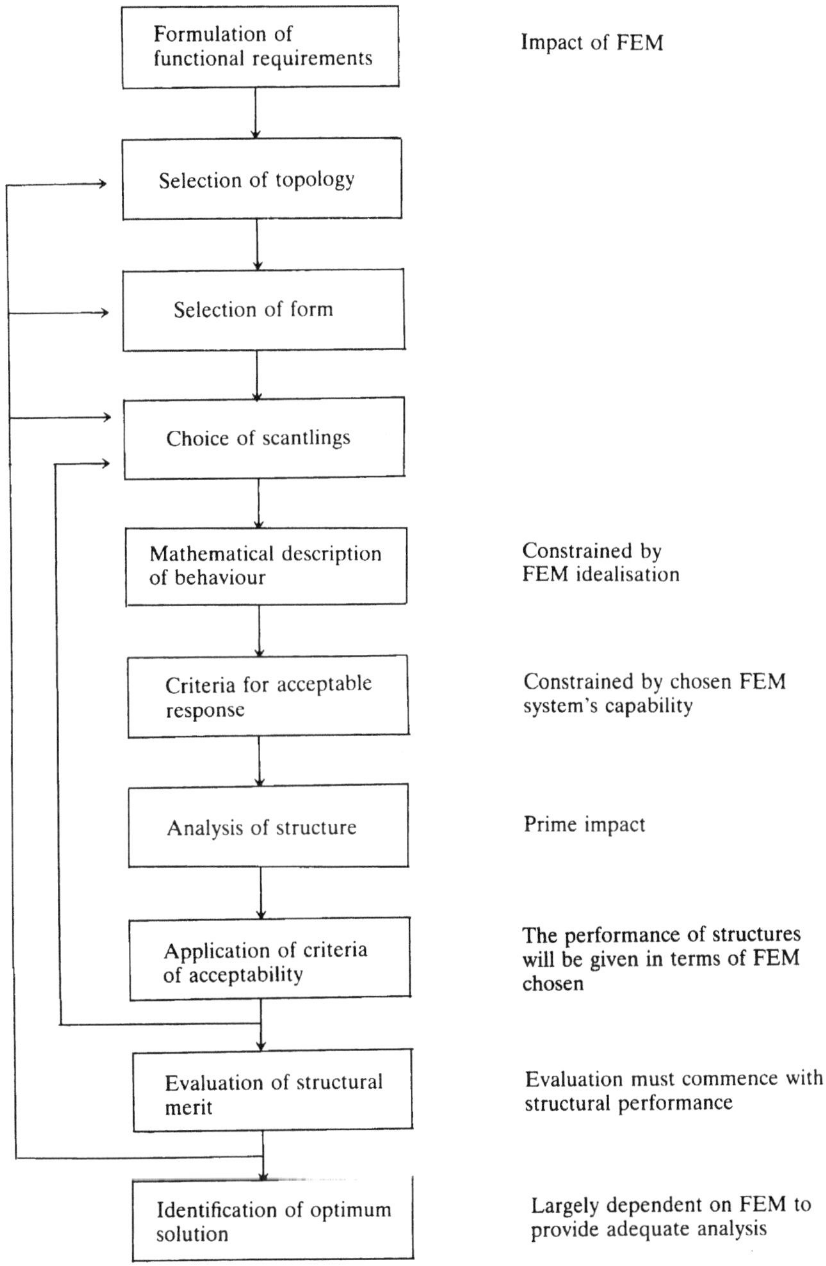

FIG. 6.1. An outline of the rational ship structural design process.

(8) *Evaluation of structural merit.* This must be related to the wider aims in the overall ship design, but is intended to be a consideration of structural efficiency, be it cost- or weight-based.
(9) *Identification of optimum solution.* Provided there is a clear basis for evaluating the merit of the selected topology and scantlings, it should be possible to carry out sensitivity studies to systematically vary those structural variables that can reveal significant improvements in structural performance.

Figure 6.1 furthermore highlights those stages in the rational ship structural design process in which the FEM plays a major role. Clearly, the finite element technique, primarily as an analytical tool, is the lynch-pin in the analysis of the structure but, in adopting such a tool the designer, consciously or not, constrains his scope in other of the steps just outlined. The comment alongside step 9 in Fig. 6.1 is significant in that all the structural optimisation approaches applied to ship structure are done as part of a sophisticated finite element system which incorporates data generation, load and structural discretization and post-processing of output.[5] The other comments show how the choice of a given FEM system can heavily constrain the other steps in the process beyond the initial choice. Perhaps this suggests that within the rational design process, as actually practiced, an important decision step omitted from this representation is that of 'Choice of analysis method', and that this would be before step 5 or be part of the decision on mathematical idealisation. The more detailed aspects involved in the choice of FEM to be adopted for ship structural analysis are discussed, after some specific examples, towards the end of the chapter.

It is suggested in the rational structural design process that in evaluating the merit of a given structure, as a direct output of the process or as part of an approach to optimising the structure, the aim is an 'efficient' structure, and that this might be seen in cost or weight terms. The concept of efficiency could therefore be seen as the aim of the ship structural design and hence, in the context of FEM applied to ship structures, it could be considered the criterion in the choice of FEM to adopt in a given instance. Although the division of ships into merchantmen and warships is, in design terms, far from straightforward,[6] the commercial objective, in the former case, provide a useful distinction in regard to structural efficiency.

If by a merchant ship we consider the transportation carrier (e.g.

bulk, cargo or oil carrier) and ignore the service vessel category, then the designer's aim of maximum structural efficiency can be considered from the wider stance of transport efficiency. In this case the easiest measure of structural efficiency—that of the lightest possible structure—will be wide of the mark. It may well be that the overall transport efficiency of a container system, of which an individual container ship is but a component, is insensitive to the ship's broad structural weight. However, the transport costs may be very sensitive to the initial cost of the structure. It has been well demonstrated by Caldwell[1] that the two aims of minimum steelweight and minimum structural cost produce quite different structural configurations. It is further worth remembering that not only is cost a less exact objective function to minimise than is weight, but also the relevant cost may be initial or procurement cost or that of running cost, where maintenance through life is involved.

In the case of warships, while transport cost is clearly inappropriate, good housekeeping has emphasised 'cost-effectiveness' and tried to relate this to the whole of a ship or a class of ships' life. However, economic constraints have recently focused attention onto the more straightforward measure of initial procurement or capital cost.[7] The overall hull structure in a typical small frigate will only contribute some 10% to the initial cost of the vessel, so the pressure to reduce the cost of a given design may be unlikely to focus heavily on the structure. However, it needs to be pointed out that the impact of a heavy (i.e. over-designed) structure on the overall design may be more dramatic where achievement of certain performance aspects such as top speed may be crucial.

In summary, the aim of ship structural design is twofold: efficiency, which may be in weight or cost terms; and reliability in structural performance, that is the avoidance of failure, which may be catastrophic or partial, i.e. leading to failure, to achieve the design objective of ship operation or economic return.

From the rational approach to ship structural design it is important that all the design aspects, and not just that of analysis, are appreciated in any choice of FEM. This is of growing relevance given the advances in CASD and the integration of some FEM systems in CASD systems. It is therefore necessary to conclude this section with a consideration of structural synthesis and, associated with that, the model of ship structure commonly adopted. The nature of design synthesis is a complex response by the designer to the potential owner's initiation. As such, the synthesis is strongly influenced by the wider design environment and the designer's own idiosyncratic input, to produce an initial model using a key generator. To be effective, this must comprehensively contain all the important features.[6] In structural terms, this 'building up of the parts to form a whole' is essentially achieved by selection of a model and the use of progressively more sophisticated checks to verify the detailed configuration and scantlings, if not the overall structural concept.

If the broad structural topology of the conventional ship is considered (Fig. 6.2), then there are basic features as well as more detailed differences between merchant ships and warships at a component level. These differences arise due to the former being designed according to classification society rules and the latter by design bureaux as the result of research into lightweight, medium- to high-strength steel structures. The latter are, consequently, far more expensive to fabricate than is acceptable in commercial practice. Given the broad commonality, in that the ship is basically a long box structure with the primary loading case, the hull girder bending in a seaway, the naval architect adopts a model such that the structure

COMMON FEATURES

(1) Very thin, walled, hollow non-prismatic box girder (side-shell, bottom and one or more decks). This is the primary hull structure.
(2) Internal main transverse bulkheads subdividing the box girder.
(3) Possibly one or more main longitudinal bulkheads contributing to longitudinal strength.
(4) Superstructures that may or may not contribute to longitudinal strength.
(5) Minor and local structure, subdividing and providing local support.

MERCHANT SHIP	WARSHIP
(1) Mild steel = 240 MN m^{-2}	(1) Higher strength steels B quality $\sigma_y = 310$ MN m^{-2}
(2) Main panel breadth (b) to thickness (t) = 500 to 3000	
(3) Single skin open stiffening by L, T, bulb plate or flat plate	(3) Symmetric T bars with slenderness 30 to 40
(4) Unsupported plate panels $b/t = 30$ to 100	(4) Slender plate panels $b/t = 60$ to 90; aspect ratio > 2·0 longitudinally

FIG. 6.2. The basic structural topology of a ship.

resisting longitudinal beam bending (i.e. about an axis amidships, at the neutral axis and in the athwartships direction $\bar{M}_{yy}$) is denoted *primary structure*. This has the advantage that the longitudinally and transversely positioned structure can be considered separately. The longitudinal loading, due to wave actions and weight distribution, predominates over the transverse loading, which causes the hull cross-section to distort. This approach of longitudinally effective structure being primary structure leads to classifying the structure in the form shown in Fig. 6.3:

(1) *Primary structure:* the hull girders resisting longitudinal bending.
(2) *Secondary structure:* stiffened plated grillages between boundaries providing portions of primary structure.
(3) *Tertiary structure:* elements of secondary structure, i.e. unstiffened plating and individual stiffeners with associated plating.

Figure 6.3 also shows these levels of stress; primary (σ_1), secondary (σ_2), and tertiary (σ_3). However, this division is purely a convenience for analysis and design. It is a simplified explanation of the stress distribution in the structural components of the hull girder at the tertiary level. Consider how load is transmitted in a long panel of unstiffened plating resisting lateral load. The load is transmitted essentially to the bounding longitudinal stiffeners with their associated plating and then to any transverse stiffeners with their associated plating; hence to the overall grillage constituting a portion of secondary structure. If the lateral load, arising from water pressure on the keel, is considered further, the load is subsequently transmitted, both by shear developed at the round of bilge to the ship's sides and from the fore and aft boundaries of a grillage in the keel, to the orthogonal grillage structure of the bounding bulkheads. This then adds to the primary load in the flanges of the box girder exhibited in the keel grillage as an in-plane load. This justifies the model of the overall stress pattern of the ship's structure as a box girder, which must have sufficient overall longitudinal modulus of rigidity. It is then possible to analyse the component grillages and their individual members.

It has to be said that such a breakdown:

(a) neglects cross coupling effects;

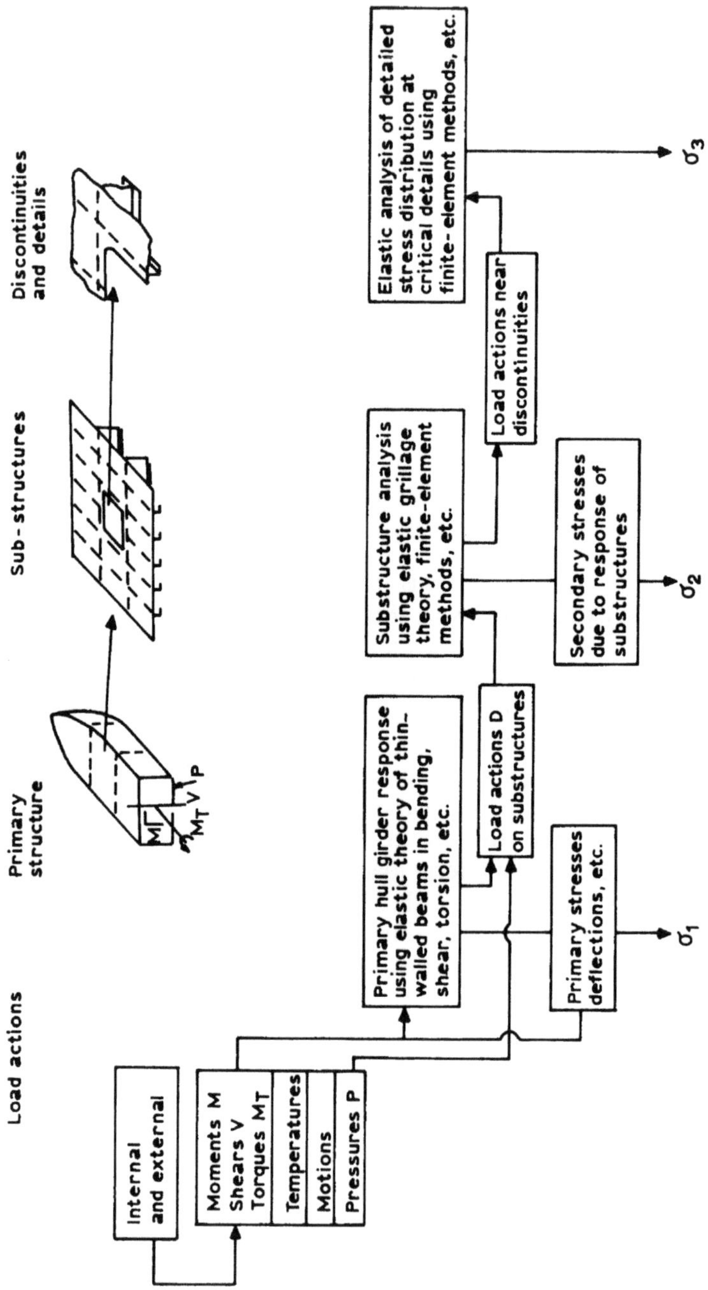

FIG. 6.3. Basic ship structural model.[1]

(b) neglects inherent non-linearities in loading and response;
(c) may not be amenable to interpreting the output of a specific finite element analysis of a portion of a ship's structure.

In designing a ship's structure, despite the advances in FEM (particularly recent developments discussed at the end of this chapter), it is generally the practice to adopt very elemental and gross expressions for the various stresses of importance, namely:

(a) predicted, i.e. direct stresses;
(b) critical, i.e. buckling stresses;
(c) ultimate, failure stresses.

Such simple expressions are easily manipulated. This is necessary while the overall ship design is fluid and undergoing many variations in the search for the customer's compromise between capability and cost, as well as the designers' convergence on a balanced and feasible solution.[6] Thus, initially approximate and non-rigorous formulae are adopted.

Normally, ship structural design starts with the synthesis of the midship section. Figure 6.4, due to Harvey Evans, indicates the component steps in such a 'section method' for a longitudinally framed structure. The approach starts from the midship section properties of the hull girder and initially neglects any other loading conditions. This is justified for the conventional monohull with certain notable exceptions, e.g. Very Large Capacity Carriers (VLCC) and container ships. The assumption that other loading conditions than that giving rise to longitudinal bending are not determining on the primary structure is usually sensible. Even if this is not the case, the approach provides an initial start point to an interactive process which checks for other loads. Although this initial synthesis is commonly used in ship design, another approach, the 'Gross Panel' method, has conceptual attractions which have become more realisable due to modern FEM systems. The ship's structure is modelled as an assembly of stiffened panels, with the location, dimensions and configuration of the panels determined by the overall ship design. Each separate panel can be designed and optimised (or strictly, sub-optimised) for known loads, after which assemblies of panels can be optimised. Modern numerical methods have made such approaches possible and the sophisticated approach by proponents of structural optimisation such as Hughes *et al.*[5] follow this broad method. However, the 'section method' is also

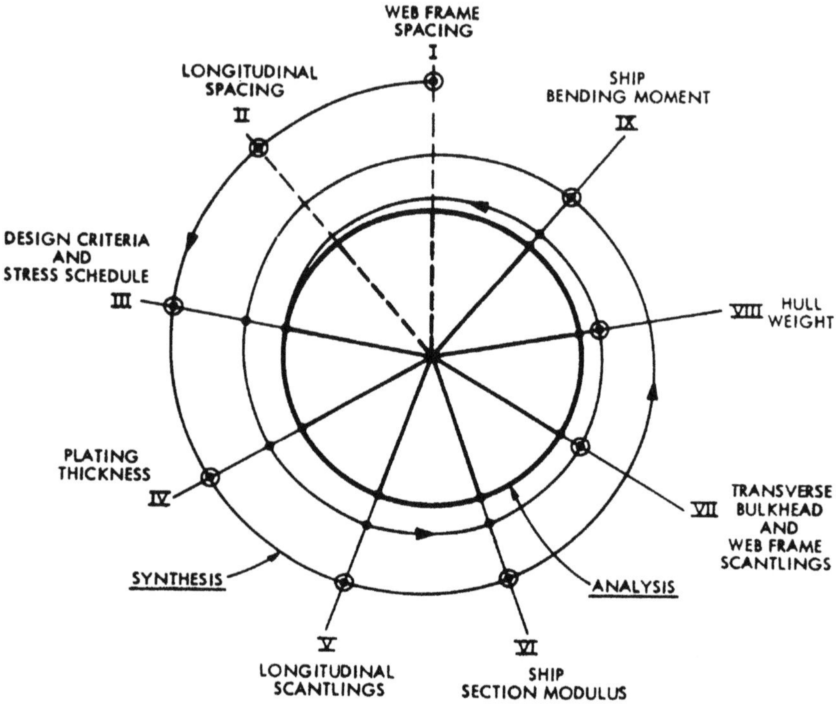

FIG. 6.4. Synthesis of the midship section.

adaptable through interactive finite element analysis of grillages to modern CASD based structural synthesis.[8]

6.3. THE NATURE OF STRUCTURAL LOADING

The actual loading of a ship is complex, and any process of structural analysis involves some degree of simplification of all the relevant aspects; loading, response, structural topology, material behaviour and failure mechanisms. It also follows that there is little point in undertaking a highly sophisticated analysis of a structure's behaviour if the loading applied is unrealistic or known only to a low level of accuracy.

The range of loads imposed on the ship can be considered in terms of the forces applied to the ship or from the sources of the applied load. In the latter case, a useful way to categorise these loads is to consider their periodicity. Thus, at one extreme, there are loads that occur once in a ship's life, during its construction and can include gravitational, thermal, wind, welding-induced, earthquake and launching loads. At the other extreme are impulsive loads which can range from regular berthing loads to full-scale collisions. These include slamming pressures set up under the hull locally due to severe ship motion. This latter instance can be severe and the whole hull girder can be forced into a vibratory response called 'whipping', which can be extreme. This example also illustrates that a simple static model may be wholly inappropriate for certain design cases, such as highly ballasted high-speed cargo vessels. Furthermore, there is the question as to how the various loads imposed on the vessel during service can be sensibly combined to represent the environment to which the structure has to respond.

As should be clear from the discussion of the normal model of the ship's structure, which emphasises the response to longitudinal bending, that it is those loads causing the hull girder to bend longitudinally which primarily determine the scantlings of the conventional ship. This longitudinal hull bending moment arises from the difference between self weight, which can vary substantially with cargo distribution, and buoyancy, which is dependent on wave action departure from the still-water case. The problem is complex in that the wave action arises from waves having different heights, frequencies and directions. If the motions of the vessel are also considered, the picture is indeed complex and hence the traditional resort to simple quasi-static models of loading, where the ship is 'balanced' on a wave (e.g. of height = $0{\cdot}603\sqrt{L}$ m where L is ship's length), have a clarity and continuing relevance in the initial synthesis. However, beyond this stage and for reasons already discussed, more realistic load representations are required. Given that the fundamental loads of interest in normal order of importance are:

(a) longitudinal and vertical, bending moment and shear force;
(b) lateral, that is transverse, bending moment and shear force;
(c) torsional bending moment;
(d) secondary loads such as hydrostatic and the displacement, velocities and acceleration responses, particularly in roll, pitch and heave.

It is now possible through strip theory[9] to predict the sea loads on a ship in regular seas. Furthermore, it is possible to represent, by means of a spectral description, the short-term sea state. To obtain the ship's response in irregular seas, it is assumed that the process of wave generation remains time independent for a short period (say 30 minutes). With the assumption of linearity, it has been shown that the response (e.g. the wave-induced bending movement) is also a stationary normal process with a zero mean. However, this powerful technique, backed up by model tests in ship tanks and measurement in service, when applied to structural response as opposed to sea-keeping, is more questionable since the structural designer is interested in conditions of extreme loading. It is precisely at extremes that assumptions of linearity in strip theory and superposition are open to question.[10,11] To combine the short-term statistical references it is necessary to consider not just the probabilistic description of the sea but also the ship's operating pattern. It is thus, at least theoretically, possible to arrive at a Long-Term Cumulative Probability (P_{L_1}) related to the typical encounter frequency of 10^8 waves over 20 years' life. The designer can obtain the probability that a given design bending moment (R) (representing the inherent capability or strength of a structure) will only be exceeded once in the ship's life, i.e.:

$$P_{L_1} = \int_0^{2\pi} \int_0^{\infty} \int_0^{\infty} \int_0^{2\pi} \int_0^{\infty} P_{S_1}(r \geqslant R, H_r, T_r, \mu_r, \chi, v) p_s(H_r, T_r, \mu_r) \times P_0(\chi, v)\, d\chi\, d\sigma\, dH_r\, d\mu_r\, dT_r \quad (6.1)$$

that is, the short-term probability weighted by the relative frequency of the respective sea states and the relative frequency of ship's speed and heading.

Where does this level of complexity leave the FEM-based structural analyst? As already pointed out, the structural designer must be aware of the inherent complexity and probabilistic nature of the actual loadings imposed on the ship. However, beyond that, the above representation of P_{L_1} relates to a limiting load. Provided that is a believable estimation, then this load can be used as an *effective design load* to be applied to the primary structure in a comprehensive finite element analysis. Before specific examples of this type of analysis are considered later in this chapter, it is necessary to say that although the outline of the theory for the above estimation of P_{L_1} is clear, there are several aspects where the theory has been shown to be inadequate.

Faulkner[11] has listed the most important of these as:

(a) Over-prediction of bending moment (B.M.) of typically 30% by linear strip theory.
(b) A lack of consistency in the over-prediction with wide scatter.
(c) Doubts as to how to convert measured strains on ships into applied bending moments because of structural slenderness, section moduli and other non-linearities.
(d) Slam-induced bending stresses (whipping), which can coincide with maximum wave-encounter sagging moment, and must be dealt with in combination with the wave-induced B.M. in a probabilistic manner.

Thus the structural designer has been forced to resort to measurements on ships at sea (which are open to interpretation) and the need to relate strains at a point in a complex structure to the overall design stress pattern. In addition, order statistics have been used to deal with prediction of extremes rather than the rigorous approach of eqn (6.1). This approach has resorted to expressions for the total bending moment of the form

$$z = m_0 + r \tag{6.2}$$

with a probability density function of z_A maximum in a record of

$$\phi_{zn}(z) = \frac{nl}{k}\left(\frac{z-m_0}{k}\right)^{l-1} e^{-[(z-m_0)/k]^l}[1 - e^{-[(z-m_0)/k]^l}]^{n-1} \tag{6.3}$$

where l and k are Weibull distribution parameters equal to 1 and λ respectively, where λ is the expected value of the random variable r. Hence the probability of exceeding an extreme wave bending moment R in n encounters is

$$P[\text{extreme amplitude} > R] = 1 - [1 - e^{-[(z-m_0)/\lambda]^l}]n \tag{6.4}$$

What has been recommended as the sensible approach, given all the uncertainties and the inability to obtain agreement between the approaches, is pragmatism. Thus, all approaches need to be applied and, based on previous representative and proven ship structures, for a given new design a believable Effective Design Load needs to be adopted, with these checks performed.

The above exposition has been fairly detailed because of the need to emphasise that in applying a given longitudinal bending moment distribution to a finite element representation of a ship structure, it is vital that the arbitrariness in absolute, if not relative, terms is appreciated by the analyst. This brings the analyst back to the essential aim of the particular analysis being undertaken. Given the degree of uncertainty in the loading, or at least the need to specify a realistic design load, this then must govern the manner in which the load idealisation does and should relate to the finite element model of the structure and its discretisation. Finally, by way of concluding the topic of loading, it needs to be emphasised that, in specific ship design, special loads such as the cargo loads on an LNG (Liquefied Natural Gas) Carrier or aircraft loads, both dynamic and static on an aircraft carrier, can be the determining load case instead of the primary longitudinal bending instance. In this regard two specific cases, which have a bearing on the utility of FEM in ship structural design, need highlighting:

(a) *Transverse loading on supertankers*. This loading case due to forces from wave action and within tanks has been one where the comprehensive use of finite element techniques in ship structural analysis has been highly successful.[12]
(b) *Fatigue*. This has been a common cause of structural failure in the sense of non-catastrophic failure. In that it relates to locally high stress regimes due to poor detailing (even though the mechanism is far from adequately predictable), FEMs are eminently suited to the detailed investigation at the secondary and tertiary structural levels. As with most of the discussion in this section, the problem of analysis focuses on the adequacy of the load representation, although in this instance it is the local stress pattern that is the problem.

6.4. STRENGTH AND SHIP STRUCTURAL SAFETY

If load is perceived as the input to the structural system comprising the ship, then strength is its capability to respond to the input, while safety can be taken as the measure of the response to the loading applied. In traditional ship design, before the 'rational' approach to ship structural design was widely adopted, safety was comprehended simply in terms of a stress limitation, and a primary stress at that; so all the designer had to achieve was a strength capability based on a simple model. This simple model and the associated stress level is still of use to the ship designer at the very earliest stage of design, but only if it is used as a comparative measure with previous proven ship structures that are comparable. This means that it is of little use once the structure of the new ship design is different from the type ship. In addition, because such an approximate model effectively deals with load and failure modes, the new ship must match the intended loading extremes and failure mechanisms of the type ship. The other major manner in which such a broad comparative approach is unsatisfactory is that it cannot give a clear indication of the degree to which the given structure, adapted on the basis of the comparative method, is over- or even under-designed. Furthermore, in reducing all the failure mechanisms to a simple stress limitation, it is not possible to distinguish the various likely modes of failure and their respective likelihoods of occurrence. Thus, adoption of such an approach has led to structures with low margins of safety arising from one specific failure mode being under-designed. It is in giving a clearer indication of the behaviour of the structure, both overall and at component level, that modern structural analysis has advanced the understanding of structural behaviour and hence contributed to a fuller picture of structural safety.

While research was being undertaken into the nature of the loading on ship structures, parallel investigations were being pursued into the strength of ships. This rapidly led to a more complex representation of the hull girder, both with regard to the inadequacies of the simple beam bending model of the hull's response and the mechanisms that lead to the failure of the hull girder in the ultimate. However, this is only part of the picture with respect to ship strength. In adopting a rational approach to ship structural design, along the lines already undertaken by the aeronautical and civil engineering professions, it was recognised firstly that the loading regime on ship structure was inadequately described by a deterministic model, due to the stochastic nature of the sea environment. Secondly, the strength of a structure is non-deterministic due to variability in material behaviour and uncertainty in strength assessment. This leads on to considerations of structural safety beyond simple factors of safety through to a fully developed theory of Structural Reliability, where a given probability of structural failure is given by the joint probability of the demand (i.e. loading) and the capability (i.e. strength). Unfortunately, as will be outlined, while this model of structural reliability has an intellectual attraction and even a philosophical validity, the inadequacy of the data to support the numerical values of probability of failure for given ship structures has meant that, for the foreseeable future, a completely probabilistic description of ship structural behaviour seems unachievable. For this reason, recourse has been made to 'second level' measures of structural safety which have the virtue of representing the significant modes of failure in a given ship structure while being useful descriptions of the ability of the structure to meet the expected demand. The role finite element techniques play in this assessment of structural behaviour is an important but necessarily limited one. The importance resides in the better description of the structure's response to a given loading regime, and how the component elements of the structure respond to the load. The proviso of this is that the finite element technique adopted must be capable of describing the relevant response mechanisms. The limitation is relevant to the measures of safety since these relate various non-deterministic and even judgmental aspects to arrive at a measure of the reliability of the structure. Thus, the contribution of a numerically strong analysis technique must be seen as vital but not overriding.

Considering the strength of the typical ship's hull girder in closer detail, in a seminal paper on this topic, Caldwell[13] showed the traditional approach to be unsound. It considered the midships section modulus (Z) to be the only measure of a ship's primary structural strength, in that the limiting direct bending stress $\hat{\sigma}_x$ is the response of the structure with cross-sectional properties I_{zz} and Z to design vertical bending moment $M(x)$ as given by Simple Beam Theory, where

$$M(x) = Z\hat{\sigma}_x$$

and

$$Z = I_{zz}/z$$

I_{zz} is the second moment of area of the midships section about the neutral axis, and z is the vertical distance from the neutral axis to the keel or upper strength deck.

Such a formulation predicts that two ships with the same Z will withstand the same applied bending moment. However, this assumes the same failure mechanism will limit the maximum bending moment, whereas quite different structural configurations susceptible to different failure modes could have the same section modulus (Z). Caldwell compared a simple idealisation, in which failure could occur only through the beam reaching yielding throughout—the plastic moment—with a similar section in which the 'outer fibre' of the half of the section in compression reaches a limiting stress less than yield due to a compressive mode of failure. This gave an Ultimate moment of resistance (M_u) less than the fully developed plastic moment (M_p) of the first idealisation. The reduction due to the compressive failure is denoted as a strength factor (ϕ) for the deck or bottom, where

$$\phi \frac{\text{total load in deck at failure}}{\text{load in deck if uniformly at yield stress}} = \frac{\text{average ultimate stress } (\sigma_u)}{\text{yield stress } (\sigma_y)}$$

The reason why Caldwell expressed the ultimate strength in generalised form is because the actual manner in which a given ship structure fails depends on the modes of failure that can occur at the secondary level of structural description. For the conventional structural configuration of a ship, the hull girder is made up between main transverse bulkheads of decks, ship's side and bottom, consisting of welded grillages of plating stiffened by structural members, either unidirectionally or both transversely and longitudinally. Given that the primary load case for most ship structures is that of longitudinal bending, this is manifested in the grillages under maximum load, i.e. amidships and at the top and bottom of the box girder (deck and keel) as uniaxial compressive loading, i.e. in the upper strength deck under hull sag and in the keel under hull hog. There are four distinct modes of failure of ship-type grillages under compressive load:

(a) plate buckling between stiffeners;
(b) interframe flexural buckling of longitudinals between frames (i.e. transverse stiffeners) in a column collapse mechanism;
(c) lateral–torsional buckling of longitudinals between frames, commonly called tripping;
(d) overall buckling of the grillage where stiffeners in both directions buckle with the plating.

The actual behaviour of grillages under compression is complex to the extent acknowledged by experts such as Faulkner, who on observing the comprehensive series of full-scale test undertaken at Dunfermline[14] over ten years ago, commented:

> "I have passed through the phase when I might have felt justified at being called an expert, until now, I scarcely dare admit that I have any real understanding".

This complexity is due to significant interactions between the failure modes which depend on factors such as physical properties and features built into the grillage, largely due to fabrication, such as residual or built-in stresses and initial distortions.

In addition to the problem of failure of the grillage components under compression, the basic strength model suffers from an additional simplification on which modern numerical analysis has been most enlightening. The basic model of ship strength and indeed the derivation of extreme bending moments from strain-gauge measurements, have made recourse to simple beam theory. In applying simple beam theory this has implied for the ship hull girders:

(a) plane sections remain plane, normal to the curve of zero stress;
(b) the structural material is homogeneous and isotropic;
(c) the deflections that occur as a result of bending are small and elastic.

Clearly, none of these assumptions apply absolutely to a real structure. The question for the structural designer is how far the actual structure departs from the predictions due to this useful theory. If a real ship structure is considered, the application of simple beam theory is further complicated by discontinuities in the hull girder due to openings and to structures which do not fully contribute to the hull girder in primary (longitudinal) bending. The classic example of the latter is the deckhouse or superstructure; that is, an erection above the upper strength deck. For superstructure design, the designer is concerned both with providing adequate strength for the superstructure, so that it can accept the strains imposed on it by the hull bending, and, where appropriate, an acceptable contribution by the superstructure to overall primary strength. When the hull bends longitudinally, compatibility at the connection to the superstructure is achieved by vertical forces in the superstructure side and in the hull's transverse bulkheads together with horizontal shear forces at the base of the superstructure extending with the upper deck (when the ship hogs). To obtain an adequate measure of the superstructure's behaviour it is necessary to take into account a large number of factors which influence its effectiveness as part of the main structure:

(a) shear lag in superstructure decks and sides as well as in the main hull deck;
(b) deck flexibility, including the effect of longitudinal bulkheads;

(c) transverse bulkhead disposition;
(d) shear deformation of the connections of the superstructure if bolted;
(e) the shape of the bending moment distribution in way of the superstructure;
(f) openings in the sides and decks of the superstructure;
(g) variation in hull and superstructure structural cross-section;
(h) hull and superstructure material, which is normally assumed to be elastic and isotropic;
(i) sides and decks of the superstructure, which may be of differing thicknesses.

These factors have been listed because they indicate the aspects which influence an important component in certain structural configurations adopted in ship design. The importance of superstructure behaviour has been such that it has been investigated in recent years using finite element techniques. In outlining some of these investigations in the remaining sections, the results can be measured against the complexity of the problem.

Before the specific applications of finite element techniques to ship structures are considered, the manner in which ship safety can be assessed needs to be outlined in a little more detail. Having said that the deterministic assessment of the strength of the hull girder is complicated, it was recognised that the quantities involved in the assessment were statistically variable. For example, there are:

(a) variations in the 'as-built' structure due to differences in material properties and scantlings, and arising from manufacturing imperfections;
(b) errors arising from assumptions made in the design, such as scaling from model tests, and limitations of both idealisation and theory applicability;
(c) variability arising from time-dependent effects, leading to deterioration due to corrosion, cracking, wear and tear and thermal stresses.

The first attempts to provide a measure of the statistical variation of ship strength were made in 1972,[15] and this was used with a probabilistic measure of long-term loading to obtain, for particular ships, values of the probability of failure (p_f), i.e.:

$$p_f = p[C < D]$$

where C is capability (i.e. strength) and D is demand (i.e. loading), assuming C and D to be uncorrelated time-invariant random variables. Thus

$$p_{\mathrm{f}} = \int_0^{\infty} \{F_C(x)\} f_D(x)\,\mathrm{d}x$$

here $F_C(x)$ is probability distribution function for the C curve, and $f_D(x)$ the probability density function for the D curve.

This constitutes a level-3 measure of structural safety in that it is able to combine all modes of failure; however, despite the attempt in 1972, a complete probabilistic formulation of neither the loading nor the strength of a ship is yet achievable. For this reason, second moment or level-2 methods are currently adopted for ship safety assessment. While level-2 methods give a scalar safety margin, in using estimates of the second moment statistical properties of the means and standard deviations of C and D they are semi-probabilistic and at least recognise the statistical nature of safety. The two methods currently adopted[17] are:

(a) The Safety Index

$$\beta_{\mathrm{f}} = \frac{\bar{C}_u - \bar{D}_e}{\sqrt{S_C^2 - S_D^2}}$$

(b) and the Partial Safety Factor

$$\gamma_0 = C_K / D_K = \gamma_C \gamma_f \gamma_m$$

where the terms are represented on Fig. 6.5.

Given that these are the formulations used to measure the safety of a given ship structure, the adoption of a given numerical technique, such as a finite element based approach, has to be seen as contributing to a more realistic representation of the structure's response to a load regime. In addition, with regard to errors arising from idealisation and theory, it can be appreciated that not only can finite element analysis provide insights, but it can also introduce additional errors. Furthermore, when it is realised that the majority of structural failures arise from human errors (blunders) which can arise in part from poor application of analysis techniques, the structural designer needs to be doubly on guard in resorting to such sophisticated techniques. While attempts have been made to deal with blunders by recourse to fuzzy mathematics,[18] this has been at a research level so far and, currently, reliance is placed on supervision and independent checking of design and construction to provide insurance against gross blunders.

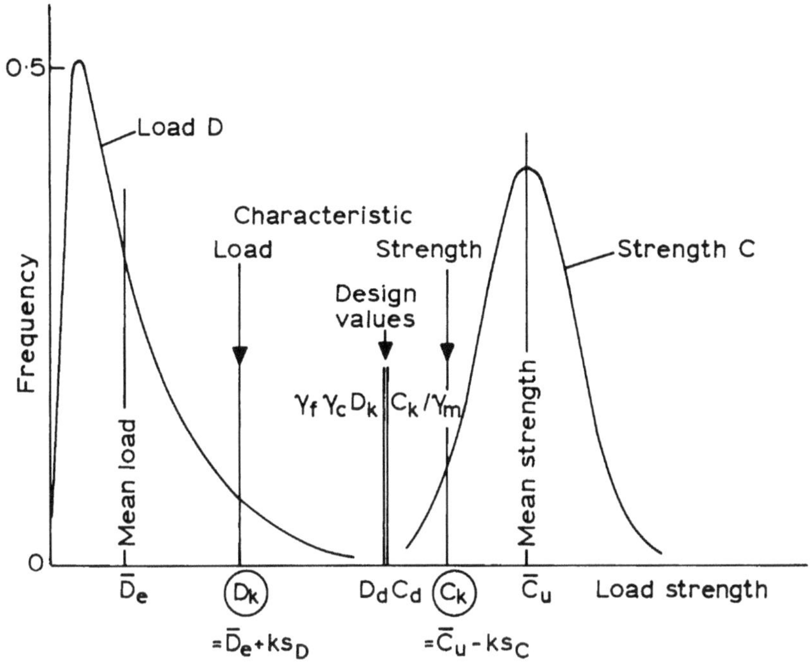

FIG. 6.5. Representation of structural safety.

$$\text{Safety Index} \quad \beta_f = \frac{\bar{C}_u - \bar{D}_e}{\sqrt{s_D^2 + s_C^2}} \leqslant \frac{1}{\sqrt{p_f}}$$

Partial Safety Factors: γ_f = force uncertainties

γ_m = material and fabrication

γ_c = socio-economic consequences

6.5. INITIAL APPLICATIONS OF THE FINITE ELEMENT METHOD TO THE ANALYSIS OF SHIP STRUCTURES

The preceding sections in this chapter have dealt with the problem of ship structural design with regard to determination of the loading and the strength. Modern analytical methods exemplified by the finite element technique have been highlighted. However, the main aim of this outline has been to provide the framework within which the application of FEM can be considered for the determination of an adequate design of structure for a typical ship. Thus the discussion of the ship structural design problem as a subset of the overall ship design task, how the former has changed with the 'Rational' approach to ship structures, together with the insights and limitations on the definition of the loadings imposed on the structure, the nature of ship strength and determination of the adequacy of ship structural safety, all place powerful analytical techniques of FEM in context. It is now necessary to be more specific as to the direct application of FEM to ship structures. In doing so, it has to be emphasised that the preceding exposition has been inevitably cursory and has, for this reason, concentrated almost exclusively on conventional monohulled steel-structured ships and the primary structural concerns in such vessels.

In outlining the application of FEM to ship structures, emphasis is placed on the historical evolution of the use of FEM in this field. This is done by drawing on several seminal expositions of such applications, as they highlight not just the nature of the task in applying such techniques to the problem of ship structures but also show vividly how there has been an intimate interplay between the development of the FEM and the understanding of the behaviour of ship structures. This then shows the manner in which FEM are used currently to assist in further understanding the nature of ship structural behaviour. The increasing incorporation of FEM in the formative design stages of a ship structure has to be seen as part of the evermore increasingly effective application of computer-aided design systems to ship design in general. The sensible way to appreciate the current situation of dual application to essentially ship structural research and ship structural design, dealt with in sections 6.7 and 6.8, is through an outline in this section of the early applications of FEM to ship structures. This outline is followed in section 6.6 by an extended example of the application of FEM in a specific major ship design, which both highlights the advantages and limitations of FEM on the grand scale, thereby reinforcing some of the FEM-related comments made in the ship structural design outline already presented.

The earliest example actually predates FEM as applied to ship structure, in that Yuille and Wilson[19] describe a matrix method approach to analysing a three-dimensional section of a frigate between main transverse bulkheads. This analysis was part of an investigation into the nature of transverse strength of a modern light-weight welded structure, given the recognition that the method used to size the

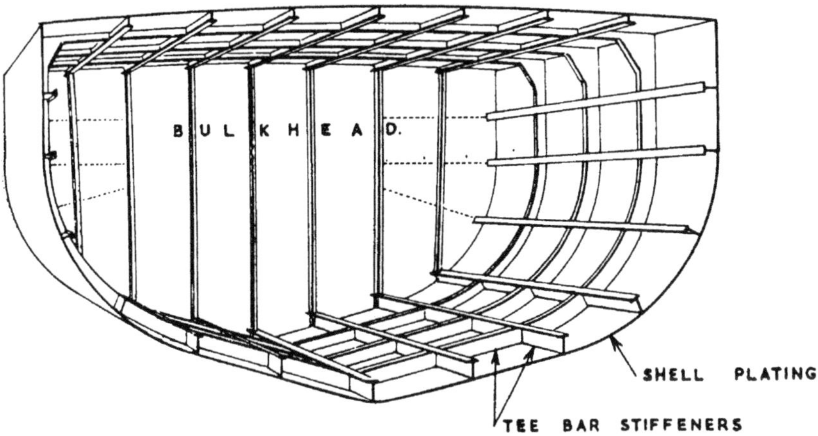

FIG. 6.6. Typical frigate structure.

transverse framing in the hull girder between main transverse bulkheads was far from satisfactory. As part of the investigation, a large (2/3rd scale) frigate section was manufactured and then subjected to loading by rubber pressure bags and hydraulic rams to simulate local hydrostatic loads and the effect of overall loading actions, both longitudinally and transversely. The matrix analysis was the only possible method to deal with the complex of curved and flat intersecting stiffened grillages (Fig. 6.6). The analytical model devised is interesting in that it had to be developed to match the measured responses in the elastic part of the test. This is partially indicated in Fig. 6.7, where it shows the need to limit the model to represent the most significant structural behaviour of the longitudinal and transverse framing as well as the plating contribution. Figure 6.8 gives the discretisation used with both the boundary conditions and the bar elements. The bar elements represent not just the stiffeners with effective areas of associated plating but also special diagonal bar elements which model the other contribution of the plating, that of shear stiffness across the diagonal. This early example is interesting in that it shows how the application of large-scale numerical methods to a structural phenomenon, transverse strength in this instance, often needs to be accompanied by a physical model test to provide a bench mark and, as in this case, to modify the analytical model to better represent reality. These are both features of FEM applied to ship

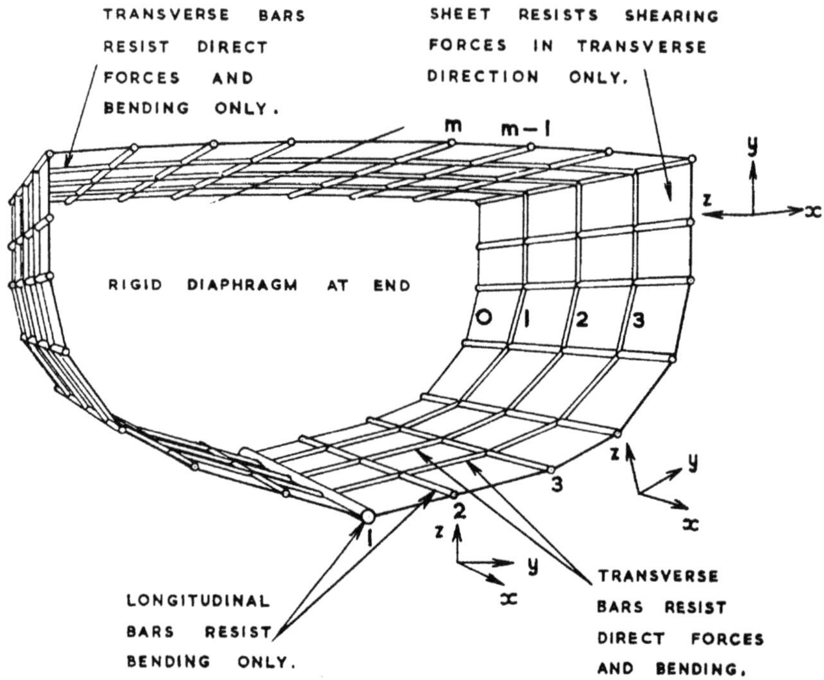

FIG. 6.7. Structural idealisation of a frigate structure.[19]

structures which have continued to occur. Clearly, this requirement for a large-scale model and tuning of the analytical model is not readily applicable in the case of routine design of ship structures, but is fed into structural design through adoption in design codes and the production of specific finite element systems as outcomes of such directed structural research.

A more representative example of finite element analysis is provided by Kendrick's presentation[12] of the investigations into the strength of supertankers in the late 1960s at the, then, Naval Construction Research Establishment at Dunfermline. Although this exercise parallels the previous one in respect of the validation of the analytical model by a physical model test (1/8th scale in this instance), it has a wider significance and lasting relevance. The significance relates to the structural problem investigated, namely the design of supertankers which had in this period changed in all respects structurally from the 20 000 tonnes deadweight post-war tankers to the monsters of

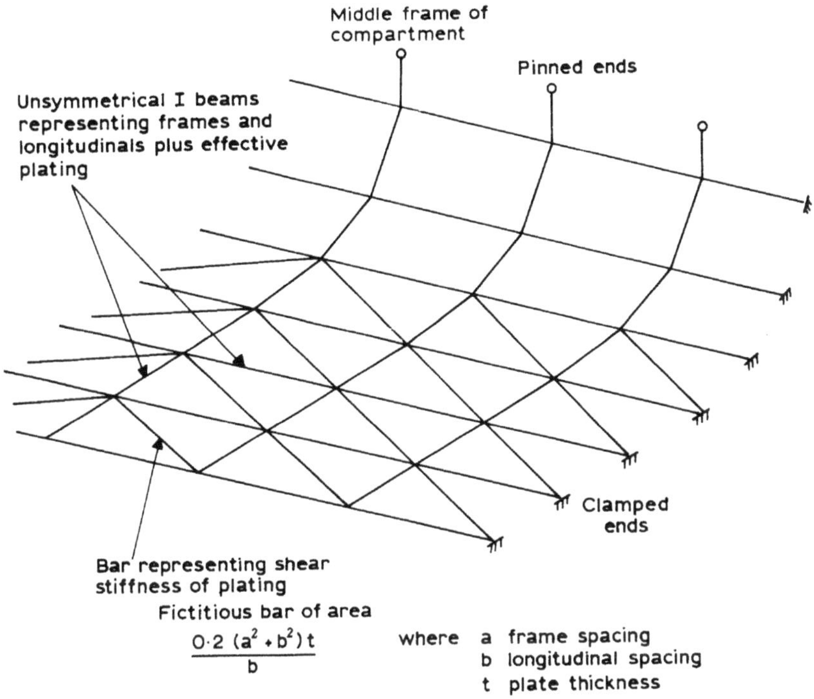

FIG. 6.8. Discretisation of a frigate structure.

200 000 tonnes and more. The problem of the structural design of such giants was no longer the primary strength consideration, discussed earlier, but rather a transverse strength problem due to increased tank lengths together with large and relatively thin cross-sections. This problem was a genuine one, as design to traditionally conservative Classification Society Rules was found to be inadequate, and it was three-dimensional stress analysis by FEM that revealed this. The lasting relevance of the exercise relates specifically to the insights produced on the application of FEM to ship structures. The use of super-elements to enable the management of the task of building up the structural idealisation of a largely repetitive structural configuration (Fig. 6.9) and the investigation of the level of discretisation (Fig. 6.10) provided insights of immediate use in applying FEM. The conclusions on the accuracy of the linear finite element analysis

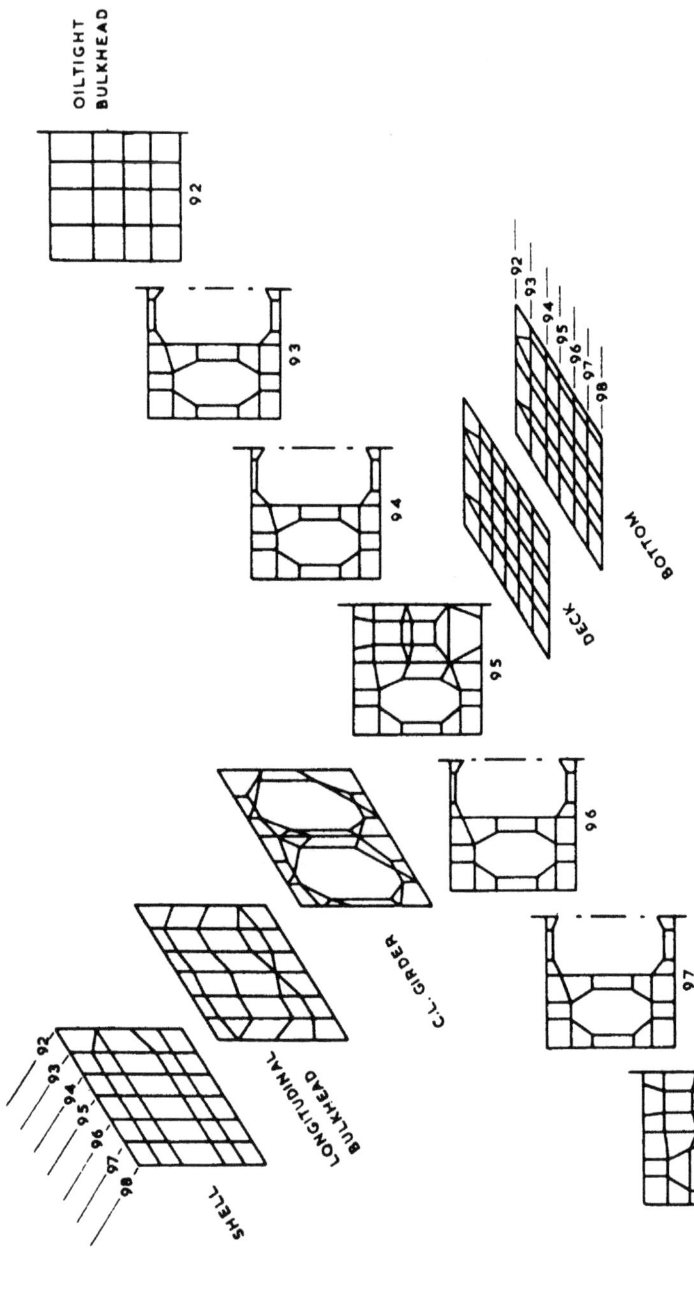

FIG. 6.9. Supertanker tank structure.[12]

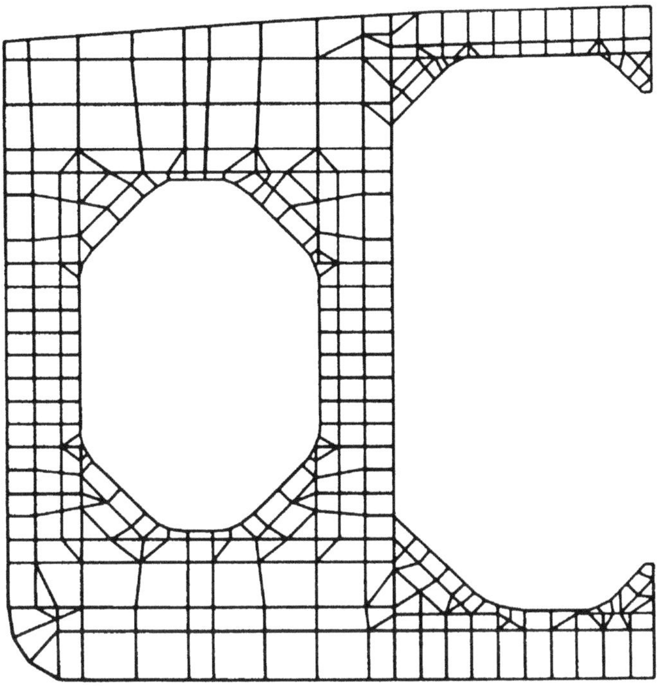

FIG. 6.10. Fine discretisation of a transverse frame.[12]

undertaken are worth reiterating, and were discussed by Kendrick:

(a) It was found that bending stiffness in the transverse frames had little effect on the plate stresses, this was fortunate as the stiffness could not be assessed to high accuracy due to the questionable basis for choosing effective breadth values.
(b) Actual material behaviour was not modelled in the structural idealisation. This included lack of flatteners in plating, as well as buckling and yielding occurring in localised areas of the structure.
(c) Questionable choice of boundary conditions selected in the model at the end of sections analysed.
(d) Finite element related aspects such as mesh size, type of elements and shape of elements selected, together with possibly poor conditioning of stiffness equations and rounding-off errors.

Finally, the analysis of what was, at this time, a novel ship structure was compared with a 1/8th scale model which was realistically loaded in a test frame. From the extensive instrumentation employed, and by varying the boundary conditions, comparison was made with the analysis and also, by loading the scale model to collapse, additional insights beyond the finite element analysis were obtained.

Two other early applications of finite element analysis to ship structures are worth briefly outlining, as each had important features relevant to the subsequent exploitation of the finite element method in ship structural design. The first of these[20] dealt with the application of finite element analysis to the problem of the hull to deckhouse or superstructure interaction outlined in section 6.4. The idealisation adopted in this case consisted of:

(a) treating the hull girder as an equivalent rectangular prismatic shell with membrane stiffness only;
(b) considering the main bulkheads and the superstructure to be flat and usually orthogonally stiffened surfaces with solely membrane stiffness;
(c) treating the bending stiffness of the supporting deck very carefully, paying particular attention to the transverse rotational restraint provided at the deck edges.

What was significant about this exercise was the fact that the analysis was backed-up by a large scale (42 feet by 8 feet) test. Five variations in bending moment distribution in way of the superstructure, along with two separate values of deck stiffness, were considered in both the physical and the analytical models. From the investigation, the importance of the assumptions on deck stiffness was emphasised. Aside from the insights into superstructure behaviour, this work, which was an early application of FEM, illustrates the benefits of having empirical verification of the use of the FEM to untried structural applications in order to gain confidence in the modelling, in terms of both structural idealisation and discretisation, adopted for a given problem. The second application[21] is an early example of combining FEM with optimisation techniques. As will be recalled, Fig. 6.1, in illustrating the rational approach to ship structural design, had as its final step 'Identification of optimum solution'. Although there is a concomitant requirement for a criterion of structural merit, it was soon appreciated that finite element analysis applied to ship structures opened up the further step of numerical analysis producing the optimal solution, within the constraints specified and according to the measures of merit specified. This example followed the approach used by most early optimisation, in that the finite element analysis was used to obtain stress levels which were interrogated for structural acceptability, while minimum weight was the measure of structural merit use. The approach to obtaining the optimal solution was based on a variation of linear programming, despite the fact that the distribution space was recognised to be essentially non-linear. That there has been a considerable development in the application of FEM to ship structural optimisation, beyond these simplified early approaches, is discussed further in section 6.8.

Alongside the essentially exploratory studies just described on the use of finite element analysis to tackle specific problems in ship structural design, it was recognised that the approach could be adopted in actual ship structural design. In doing so, structural designers were faced with two alternatives, either to adopt existing finite element packages that already existed in the civil engineering and aerospace fields or to develop specific suites of finite element programs for analysing ship structures. While both approaches have been used,[22,23] the following section outlines at some length a specific example of finite element analysis as part of an overall ship design, where the suite used was one developed for supertankers but, in this case, is applied to a more complex structural configuration.

6.6. THE USE OF A LARGE-SCALE FINITE ELEMENT ANALYSIS IN A SPECIFIC SHIP DESIGN

The previous section outlined the evolution of FEM applied to the analysis of ship structures. This application was both a proving and adaptation of the finite element approach to ship structures and the use of a powerful new technique to deal with specific structural design problems. In order to bring together many of the issues already identified in this chapter, a specific example of the use of FEM for the large-scale structural analysis of an actual ship design is now outlined. The design is that of the *Invincible* Class aircraft carrier built for the Royal Navy in the 1970s and described in detail in ref. 24. As stated in section 6.2, the structural design must be seen as a component element of the total design task. Furthermore, the design task and the emergent design are the outcomes of the role intended for the ship and the constraints on the design evolution. In the specific case of the *Invincible,* the design evolved from studies of command cruisers to replace the interim command ships following the demise of the Royal Navy's post-war new-construction fixed wing aircraft carrier design known as CVA-01 in 1966. This design, which was over 50 000 tonnes displacement and catered for Mach 2 aircraft with all up weights in excess of 50 000 lbf proved too expensive to proceed with. While this design could be seen as a logical extension of fixed-wing aircraft carriers that then existed, the emergent design of the *Invincible* carrier, although considerably smaller at 19 500 tonnes displacement, was radically different from previous designs.

Reference 24 lists the reasons why warships generally have become space-dominated as opposed to weight-dominated, the case prior to the electronic age. In this respect the *Invincible* Class design, as the only major surface warship designed and built for the Royal Navy since the war, is a voluminous ship of modest displacement. Probably the major factor in reducing the mass of the vessel while the volumetric aspect increased was the adoption of modern structural design already prevalent in the much smaller frigates and destroyers in the fleet. However, additionally, the structural design had to cope with other demands which arose from the novelty of the design, for example:

(a) The large hangar, both long and high due to the large helicopters and the maritime VSTOL (vertical take-off and landing) aircraft, and situated immediately under the flight deck, thus requiring large unsupported spans.
(b) The need for a through passage deck immediately below the hangar on which all the large communal compartments are situated.
(c) The adoption of gas turbine propulsion machinery requiring a large volume for airflow and vertical access from below the hangar and around its sides. In particular, such down- and uptakes lead to the structure being pierced by large openings in the decks and ship's side.
(d) The decision to optimise ship availability by facilitating exchange of major equipment items rather than *in situ* maintenance led to further large openings in the decks and in the longitudinal bulkheads of the hangar.

The resultant general arrangement is summarised by Fig. 6.11, which

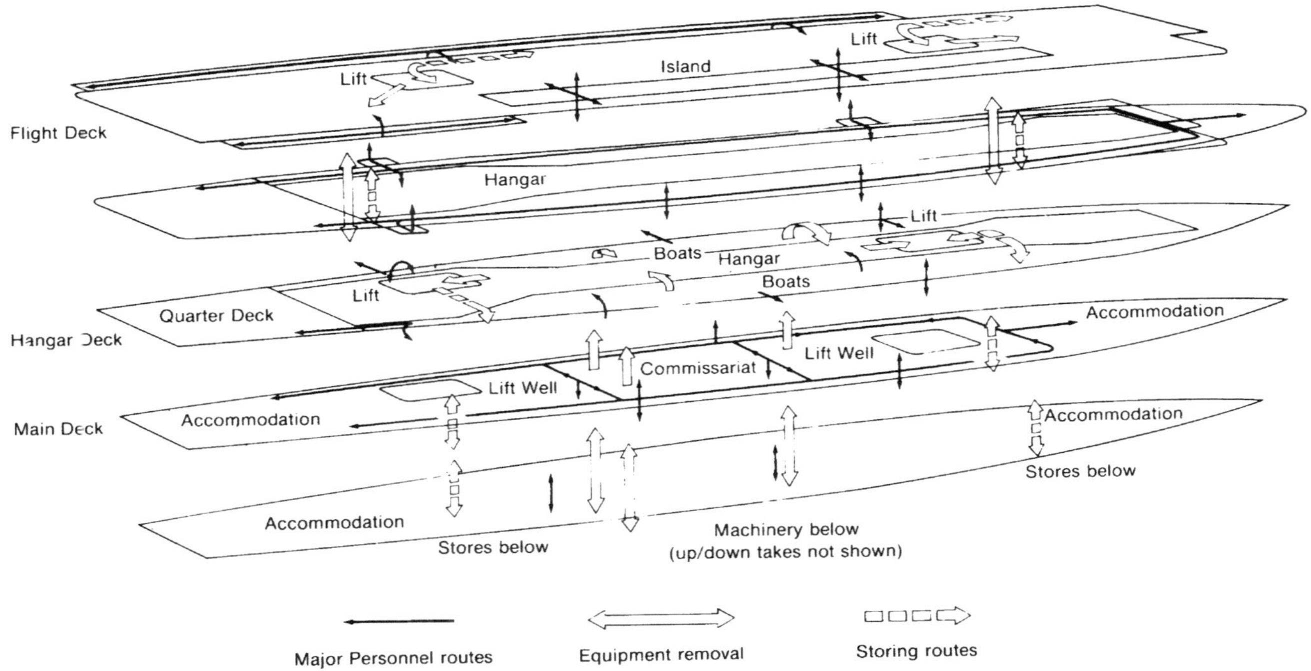

FIG. 6.11. *Invincible:* general arrangement aspects.[24]

shows not only the long hangar but also its 'dumb-bell' shape, arising from the service ducts described at (c) and (d) above. Also apparent are the two large aircraft lifts which further complicate the span problem on the upper deck, together with the extremely long superstructure which, by virtue of its length, cannot be neglected in consideration of longitudinal continuity despite its narrowness. Thus it can be seen that general features of this novel design had significant impact on the structural configuration that made the latter a far from straightforward structural design problem. In considering how the finite element analysis was applied to the *Invincible* design, it is sensible to first consider the structural design philosophy, then why and how the finite element analysis was attempted, consider typical outputs, and conclude with the general design lessons learnt from this very comprehensive exercise.

Figure 6.12 gives a simplified midship section with the principal loads that were significant in sizing the ship's structure. At the midship section the hangar longitudinal bulkheads provide a noticeable degree of longitudinal strength; however, they are not continuous through the

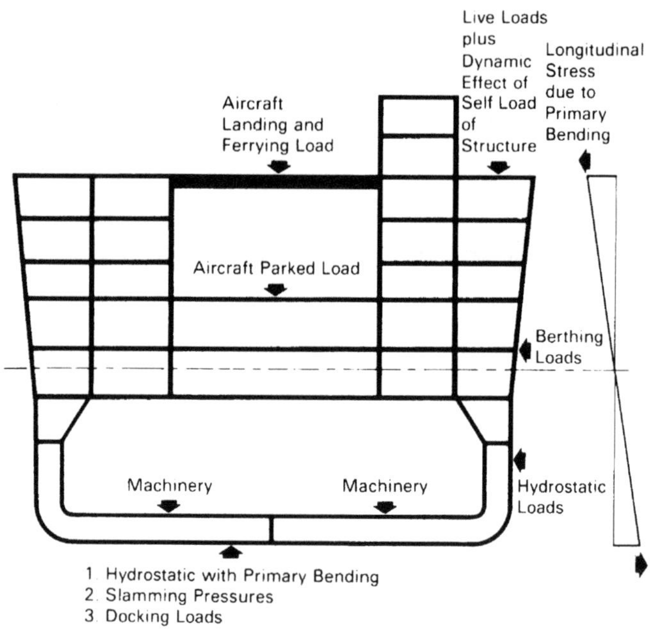

FIG. 6.12. *Invincible:* structural philosophy.[24]

depth of the structural section into the main machinery spaces. Thus, the structural configuration is unlike both previous aircraft carriers and, with the large hangar space, also unlike cruiser arrangements. This novelty was a prime reason for resorting to a large-scale finite element analysis. Before this analysis was arrived at, the structure had to be built up, and Fig. 6.12 also indicates the principal loads:

(a) *Longitudinal bending.* This was tackled in the traditional manner by initially using an empirical formula derived from the nearest comparable design, modified by *Invincible*'s form parameters and with judicious margins for the large uncertainties. As the design progressed a standard longitudinal balance on the traditional L/20 trochoidal wave was employed, using the weight distribution.
(b) *Aircraft impact loads.* The local structure on the upper deck had also to be designed to overcome aircraft impact, and this required consideration of possible future aircraft with increased weights and speeds on landing.
(c) *Live loads.* For the weather decks the dynamic effect of parked aircraft and vehicles as well as the load due to snow and ice had to be used.
(d) *Hydrostatic loads.* Loads due to seas on weather decks, sides and bottom structure, including slamming pressures, were applied.

For individual structural features particular loads and combinations were considered. The choice of practical structural scantlings demanded prudent judgement on likely combinations of loads. Based on past practice, factors of safety under specific loading regimes were chosen for important elements of the structure, for example:

(a) main deck and keel direct stresses under longitudinal hogging and sagging;
(b) main deck and keel shear forces under longitudinal hogging and sagging;
(c) secondary structure subject to direct stress, average shear stress, buckling of webs under combined bending and shear, overall buckling of stiffened panels under primary stresses, buckling of panels between frames, and buckling of unstiffened panels of plating.

It was clear by about late 1970, when the structural design was sufficiently advanced, that it would be highly informative to undertake a comprehensive analysis of the structure, given the novelty of the structural configuration and the application of modern structural scantlings to such a large structure. The specific areas of uncertainty that were identified as requiring further analytical assurance were:

(a) The stress in way of groups of closely spaced openings in major structural members, i.e. sides, decks and longitudinal bulkheads.
(b) The effect of transverse asymmetry of the structural form on the stress distribution. This asymmetry arose from the need to balance the large transverse off-centre weight due to the long superstructure with above water hull flare on the port side.
(c) The requirement for pillars to support the double bottom against external loads, especially those caused by docking (see Fig. 6.12).
(d) The contribution made by the longitudinal hangar bulkheads to primary strength.

This need for an analysis which looked at the behaviour of the ship's structure coincided with the development of the finite element techniques of ship structures to a workable level, as has already been described. It was decided to use the SESAM-69 system developed by Det Norske Veritas to carry out a three-dimensional stiffened membrane finite element analysis of the majority of *Invincible*'s structure.

SESAM was selected because of its multilevel super-element capability, and the structural model produced represented 85% of *Invincible*'s structure. The model finally comprised 34 000 elements, and thus the use of super-elements (SELs) was essential to manage the data preparation. The SEL, as a generalised substructure of arbitrary shape and size with the ability to accept any spatial variation of loads and constraints together with any number of retained or 'super nodes', enabled the structure to be dealt with hierarchically. Thus some 90 first-level elements were combined into four assemblies, which were in turn combined into a single assembly containing 220 nodes (Fig. 6.13). Five static load cases were analysed:

(1) a longitudinal bending moment;
(2) a vertical shear force applied at the forward end;
(3) a design hogging condition;
(4) a design sagging condition;
(5) a design docking condition.

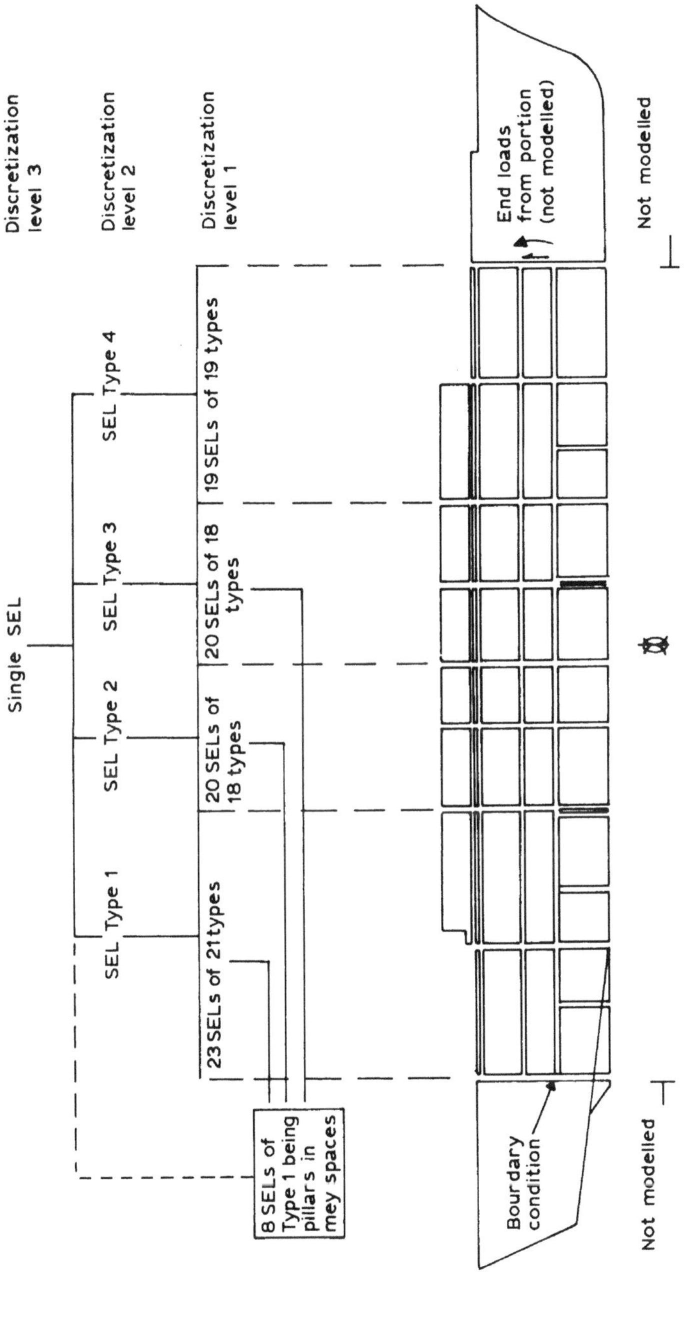

FIG. 6.13. *Invincible:* finite element analysis super-element breakdown.

Of these, the first two were simple checks on the overall response of the structural model and were compared with the Euler–Bernoulli beam response, while for the latter three cases, the loads were distributed in a realistic manner throughout each transverse section of the model. Since it was overall structural response that was of primary interest, emphasis was placed on the model's response to in-plane loads. This response is determined solely by translational displacements, and in many regions only two displacement components were considered significant. Thus, at many nodes, only two degrees of freedom were required unlike the five or six required for a full bending action response to local loads. This had the advantage of reducing the numerical computation and meant that the 14 000 nodes led to some 39 000 degrees of freedom.

Figure 6.14 presents the discretisation of a transverse section of the *Invincible* structure between two transverse stiffeners in a position near amidships for the starboard half of the hull. It can, therefore, be readily compared with the representation or idealisation of the structural midship section given in Fig. 6.12. It can be seen that the representation uses five super-elements and the supernodes which are retained in the next level of the finite element model. The figure also shows the basic elements in the model; that is, those elements at the lowest level of the finite element model, representing both the plating and the stiffening. Just how the stiffeners are represented in a specific super-element that has been extracted from the overall model, once that had been loaded and the resultant loads and deflections applied at the super-element boundaries, is shown in Fig. 6.15. This gives the discretisation adopted in the model from the structure shown in Fig. 6.16. This portion of the upper or flight deck was of major concern due to the large discontinuity presented by the aircraft lift opening in the flight deck to raise aircraft from the hangar. The discontinuity in the flight deck is compounded by several other openings in the deck structure. The need to compensate for the structural material lost was foreseen, and achieved by use of thicker plating and heavier local stiffening as shown. What the finite element analysis provided at this level was a truer representation of the local loading regime by application at the supernodes of this super-element coupled with refined discretisation. It was not necessary to apply this refinement to the overall structural model, and thus an efficient and representative model of a specific area of interest could be obtained. It was also possible to reconfigure the local structure to explore the effect on the

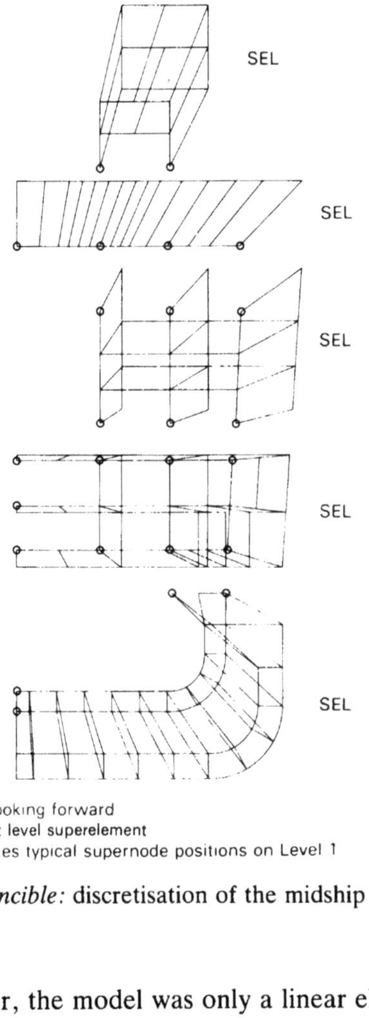

FIG. 6.14. *Invincible:* discretisation of the midship section.[24]

stress regime; however, the model was only a linear elastic one and so the investigation was limited in the insight provided compared with the complex structural behaviour likely in reality.

Figure 6.17 summarises the variation in longitudinal stress in the starboard side of the midship section of the structure from the first-level model represented by Fig. 6.14. This distribution of longitudinal stress through the decks, together with the displacement in the starboard side of the hull, is for the design hogging condition at (c) above. Several significant features were revealed from the analysis, in

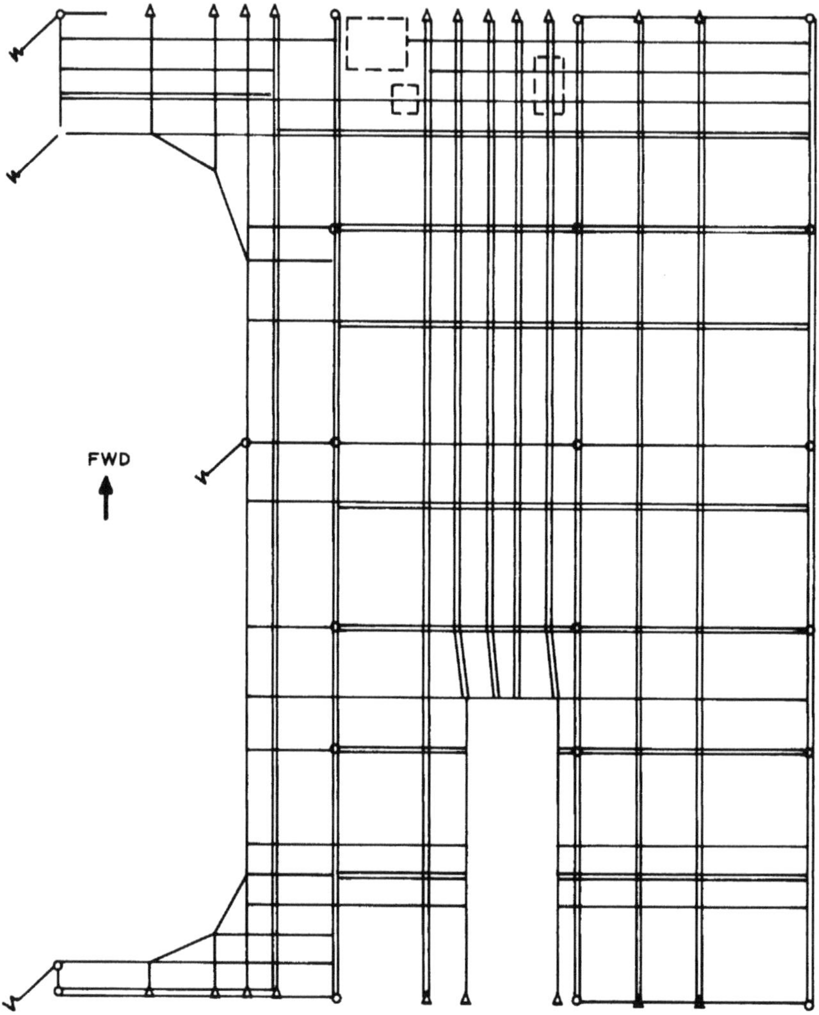

FIG. 6.15. *Invincible:* discretisation of a portion of flight deck.

particular:

(a) The marked shear lag effect across the section of the hull. While this was to be expected, it was complicated, in this case, by the pronounced local minimum shown in way of the island or superstructure. This local distortion was probably due to the sizeable deck opening just forward of this section.

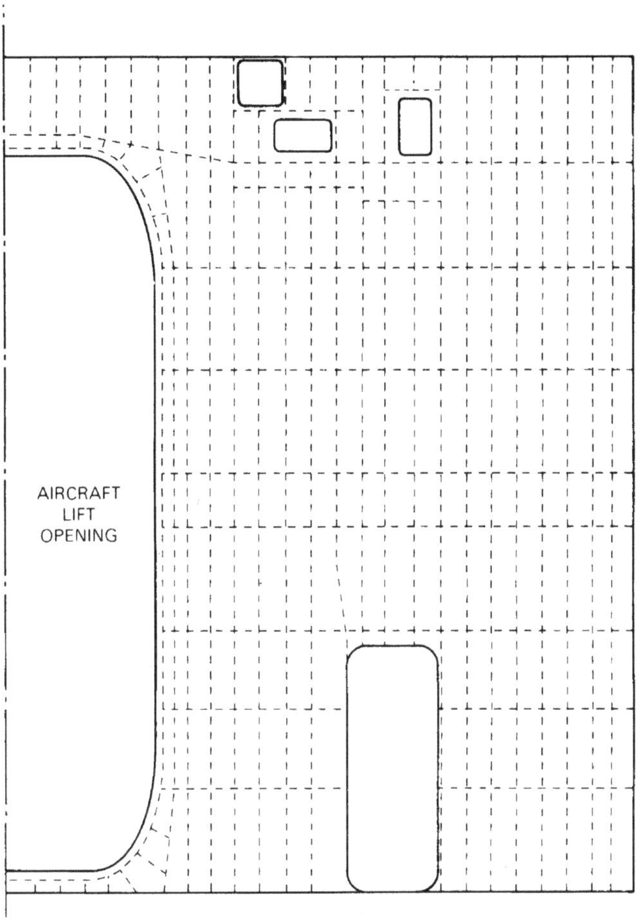

FIG. 6.16. *Invincible:* a portion of flight deck structure.[24]

(b) The plot of the neutral stress curve, or neutral axis, inboard of the longitudinal bulkheads forming the hangar boundaries. This had the unexpected consequence of causing relatively high stresses in the lowest continuous deck which had only been designed to take the loading of equipment and self weight.

(c) The high levels of primary stress in the flight deck, superstructure sides and in the double bottom structure.

As already mentioned, the overall model was needed to enable

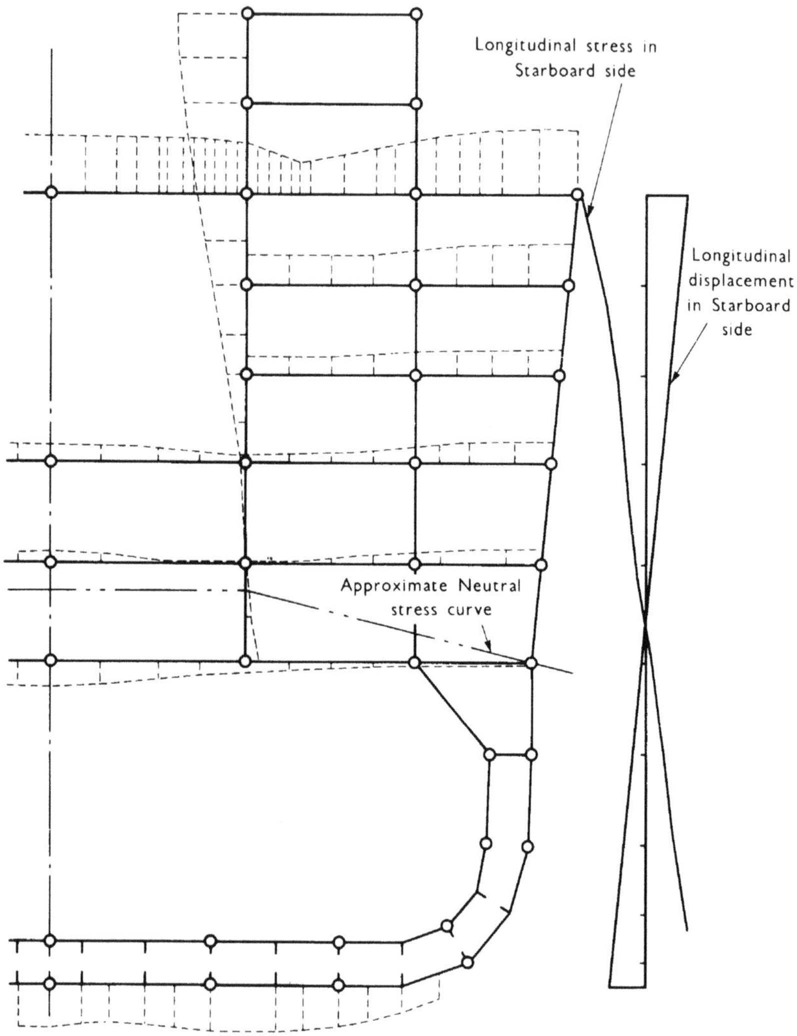

FIG. 6.17. *Invincible:* summary of finite element analysis output stress midships.[24]

detailed analysis of specific areas of structural concern, such as the flight deck. Figure 6.18 shows the stress pattern obtained for the portion of the flight deck discretised in Fig. 6.16. The pattern is indicated by isostress contours, although at the time the analysis was produced this overview of the stress regime could only be obtained

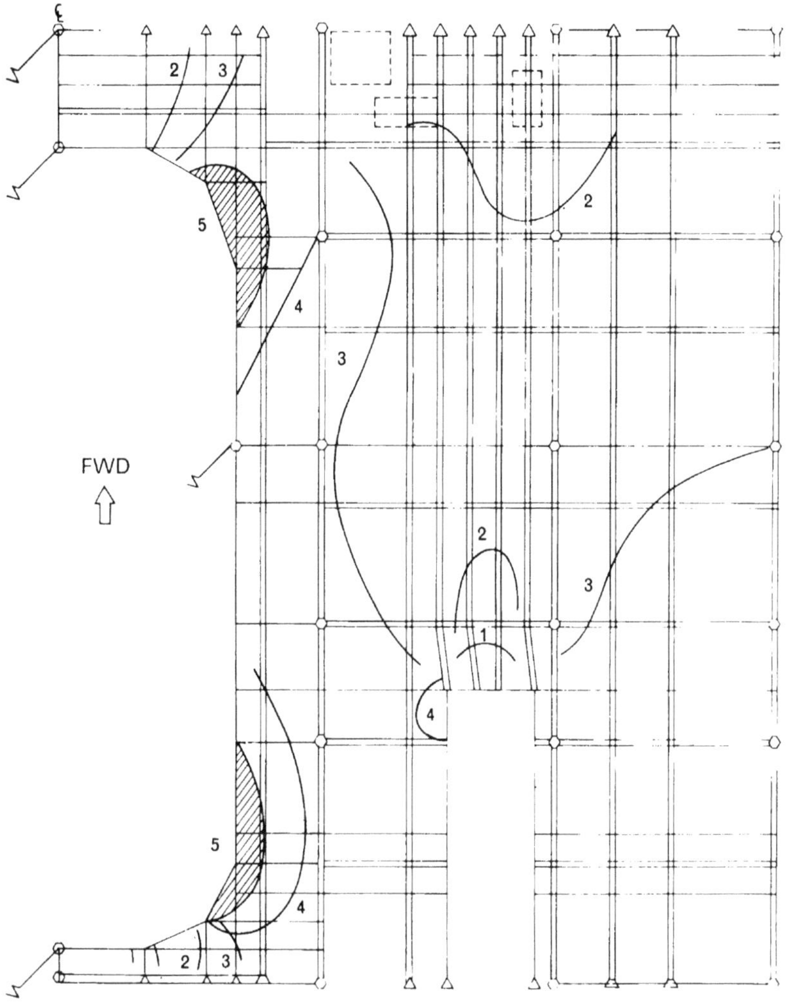

FIG. 6.18. *Invincible:* longitudinal stress distribution, part of flight deck.[24]

laboriously from the printout. The pattern, which is only indicated by relative values as detailed figures were not released, is that that resulted from the final model produced by modifying the structural configuration to eliminate unacceptable stress levels revealed by the previous analysis. As would be expected, the resultant pattern shows highest stresses in the 'hatched' regions of the lift opening corners, where local detailing was undertaken to avoid susceptibility to fatigue.

Considering the specific areas of uncertainty listed earlier that led to this particular large-scale finite element analysis, the conclusions from the use of the analysis to these specific concerns were as follows:

(a) Pronounced interactions were found for the stress fields obtained in way of openings in the superstructure area of the flight deck, and also in way of ship's side openings where the latter were of unequal height, i.e. single and two deck high openings for access and boat davits. As a consequence, in both cases the structure was modified locally and the modified structure re-analysed to determine the adequacy of the final configuration. Other areas with openings were found to be less critical and so traditional design was proven adequate in these instances.
(b) It was not found possible to separate the effect of the transverse asymmetry of the structure from other aspects such as local and general shear lag and the other major structural discontinuities; however, it was clear that there was a significant departure from the predictions of simple beam theory.
(c) The suspected requirement for pillaring in the machinery spaces to maintain continuity of the longitudinal bulkheads was not demonstrated. Because the length to width ratio of the machinery compartments was relatively low, coupled with the limited numbers of pillars possible and the low decking loads, the need for pillars was shown by the model to be questionable.
(d) The distribution of shear stress in the hangar bulkheads closely followed that of the static balance or primary shear force, which indicated that these bulkheads contributed substantially to the primary strengths of the hull girder. The main regions where the stress departed from this pattern were localised to the ends of the island superstructure and in way of large cutouts in the bulkheads for removal routes and access.

Finally, and less expectedly, the analysis also showed that the longitudinal stress in the superstructure was of a high level.

There were three major conclusions from this large-scale analysis of an extremely complicated structure:

(1) Many of the assumptions that have to be made in the traditional

approach to ship structural design, in order to synthesise the structure, are not strictly valid when it is possible to undertake a large-scale and detailed analysis of the whole structure.

(2) If the capability of the finite element method to analyse the response of the complex three-dimensional structure of a ship is to be properly exploited, then it is necessary to use the analysis in combination with a truer depiction of the structural environment and a fuller understanding of the way the structure is subject to loading, together with the manner in which it eventually fails.

(3) An analysis of the scale undertaken, with some 680 000 items of structural information, at the time it was undertaken, without the benefit of pre-processors for data preparation and sophisticated computer graphics for post-processing the results, was a complex and lengthy task demanding considerable time and effort from staff who were highly knowledgeable in ship structural design.

6.7. THE APPLICATION OF FEM TO THE ANALYSIS OF SHIP STRUCTURES

It is apparent from the early examples of the application of finite element analysis to ship structures that the method provided insights into the behaviour of ship structures that had not previously been possible. Although the early applications were largely intended as preliminary approaches to assist the structural designer in providing a more sophisticated representation of the structure of a ship and the way that structure responded to the loads imposed upon it, it was realised that large-scale analysis at this stage of development was not able to be undertaken early in a new design. Thus, the structure of *Invincible* as just described was determined by classical means, and the subsequent analysis using SESAM was a check on the design rather than a means to design the structure *ab initio*. The large effort required to set up a model of such a ship structure, validate the model, undertake the analysis and subsequently interrogate the considerable output was only seen to be justified in cases where the novelty of the structure meant that, even at a late stage in the design, there remained uncertainty with respect to the structural design produced. However, what was also clear from such investigations and the parallel developments in ship's structural knowledge into the environment, the loadings imposed on ships, modes of structural failure and concepts of structural safety, was that the general technique of FEM could and should be applied as a research tool in the field of ship structures. Furthermore, the tool could be applied to gain insight into the behaviour of ship structures at both a macro and a component level.

In considering the behaviour of the overall ship structure, the FEM enables the structural researcher to proceed beyond the simple hull girder model with its necessary assumptions about the contribution made to longitudinal bending by the superstructure and, for example, the partitioning of the response into longitudinal and transverse models of structural behaviour. Considering the specific problem of superstructure contribution or effectiveness in the longitudinal bending of a ship's hull, this had been easily recognised as a problem well suited to FEM in developing the finite element approach to ship structures. Specifically, in the *Invincible* analysis, the stress regime in way of the superstructure, despite the small transverse extent of the superstructure in that case, was shown to be high, although this would not have been obvious from classical analysis. While in this case this insight was essentially relevant at a detailed design level, the contribution of the superstructures to hull girder strength in smaller warships has been shown to be critical[15] and so the problem has been studied specifically from a research stance. The investigation reported by McVee[25] appreciated that the problem of superstructure effectiveness was related to the specific configuration of the superstructure and the hull structure in way of the superstructure. It was therefore decided, unlike the earlier American investigation[20] to model an actual ship's structure to the extent of representing the actual bulkhead disposition in the hull under the superstructure rather than using some devised deck flexibility. The detailed arrangement of the superstructure was also modelled using SESAM and the mathematical model was validated by a Darvic model. The important insights obtained from this analysis related both to deflections and the stress distribution. Deflections obtained in the superstructure were high in the superstructure, peaking when the hull traverse bulkheads were well apart while the superstructure bulkheads were closely spaced. The stress pattern exhibited large shear lag, peaking at mid-length of the superstructure on the superstructure's lowest deck with the stress in the second superstructure deck being negligible. This analysis indicated clearly that specific configuration aspects were crucial in determining a superstructure's effectiveness, although a relatively gross model could provide the designer with considerable insight.

The aspect of transverse strength has already been mentioned as a specific problem where earlier forms of structural analysis had inadequately modelled the ship's behaviour, especially when curved orthogonally stiffened grillages between main transverse bulkheads were considered. The early matrix analysis[19] already outlined had the advantage that the model could be tuned to reflect the response obtained from the large-scale model of a ship's structure subjected to loading in a test frame. While this was justified for this particular research, the sensitivity of the mathematical model to the values of effective breadth of shell plating, acting with the stiffening in bending, in the case of the matrix framework model, meant it was of limited guidance to subsequent finite element analysts. However, using a sophisticated finite element analysis suite,[26] it was possible to model the plating and framing without invoking assumptions on the contribution of the plating to the stiffening. From plots of deflection and of direct and shear stress, it was possible to explore the various features that contribute to the strength of such structures under transverse loadings. From a design point of view such an approach is likely to be considered a luxury; however, the insights from such researches can be useful in informing designers of the major determinants of structural response and the locations of prime candidate regions of structural failure.

It has long been recognised that the substructuring facility of finite element systems enables the structural designer to home in on specific structural features and, from the macro response, provide realistic boundary conditions for specific components such as bracketed connections and openings. However, it is also the case that detailed analysis can also be used to give insight into failure mechanisms. So far, the finite element analysis discussed has been applied in a linear elastic material response manner, despite the fact that it is non-linear and particularly instability modes to which ship structures are primarily designed. While the computational task in analysing a ship's structure is massive, in the case of a whole ship analysis such as that described above for *Invincible,* the task of undertaking such an analysis with non-linear element response would be extremely daunting. The approach used on non-linear instability failure is essentially one of component analysis. Thus, a stiffener or panel of plate is defined by a series of special elements, the behaviour of which is structured to represent not just the instability response due to loading but often the pre-loaded imperfections of the structural component. The latter complication is often required since the behaviour of stiffened welded grillages in instability modes of failure is sensitive to initial imperfections of shape and of built-in or residual stresses. Thus, for example, finite element analysis has been used alongside structural testing to investigate behaviour of grillage stiffeners[27] and of panel or flight deck plating[28] to produce design guidance. The immense advantage that finite element analysis has given structural researchers is that, alongside a few structural tests, a systematic variation of a range of parameters can be undertaken. These then give the structural designers clear insights into the structural impact of the choice of configuration and scantlings for special applications, such as the effect of aircraft impact on the flight decks of ships and offshore structures.[28]

6.8. THE APPLICATION OF THE FINITE ELEMENT TECHNIQUE TO SHIP STRUCTURAL DESIGN

The previous three sections of this chaper dealt with the application of finite element analysis to ship structures. Both in the case of analysis in order to advance knowledge of the behaviour of ship structures, and in the case of large-scale analysis of the structural design of a given ship, such as *Invincible,* these powerful methods are being used after the event rather than as tools the designer can use to assist him in the act of designing a ship's structure. This could be thought to be inevitable given the nature of the FEM, which requires a precise definition of a given structure in order to discretise and then analyse it. While this remains the essence of the FEM, there have been two parallel developments that have recently enabled the ship structural designer to use the powerful capability of the FEM for structural analysis while the designer is still evolving the design. Thus the finite element techniques employed during the preliminary design of ship structures can be considered as part of the synthesis stage of the design, assisting the designer not just in scantling selection but in configurational options. The two developments are:

(a) The production of computer-aided ship design (CASD) systems that incorporate structural design as an integral component of the ship design description, auditing of the design, and analysis facilities across the range of ship design features. A prime example of such a capable CASD system is the Ministry of Defence GODDESS system.[8,29]

(b) The incorporation in finite element packages of pre-processor programs that enable the designer to quickly check the model of a tentative structural solution, together with post-processing facilities that provide the structural designer, at a glance, with a clear overview of the structure's likely performance.

By linking the latter into the former, the structural designer is presented, for the first time, with a capable analysis of a ship's structure within an evolving overall ship design. The further logical development suggested by Fig. 6.1, namely to build in a capability to carry out optimisation of the structure, has been demonstrated by Hughes *et al.*[5] in the SHIPOPT system, which provides an interesting contrast to the structural capability of the GODDESS approach.

Beyond the initial or conceptual stage of a ship design[6] when, in the case of a merchant ship, a contractual proposal is produced or, for a warship, an initial exploration of the sensitivity of the design requirement is undertaken, the design begins to fair up and the naval architectural features are derived. Early CASD program suites automated the initial phase of ship design by reiterating using the simple algorithms sufficient to describe the formulative stages of the design. In such concept or preliminary CASD programs, structural aspects hardly figured in the design and this is still largely the case except for the more advanced or unconventional vehicles such as small water plane twin-hull (SWATH) vessels where, just to obtain a crude sizing, considerable definition of the structure is necessary.[30] For the conventional monohull, the preliminary sizing can be adequately performed by simple weight fractions for the steelwork; however, suggestions have been made to incorporate more sophisticated estimates of steelweight within preliminary design.[8] The most complete suggestion deals with the primary structural weight by resorting to the weight distribution, of which the primary structure is a significant component.[31] However, even at this level of sophistication, FEM is unlikely to be involved until well into feasibility design. It is at this subsequent stage that it becomes necessary to undertake firstly primary structural design and then design of the main grillages in the hull girder in parallel with the determination of the other S^5 aspects of the design to assure the feasibility or viability of the selected design option. Thus CASD systems tackling this comprehensive phase of the design need to be both sophisticated in their description of the ship and subtle in the manner in which the various analytical programs are accessed within the CASD system as the design evolves. Within this broad framework[29] the structural element is no longer just concerned with determining, with great accuracy, the weight and centroidal contribution of the structure to the overall design: it is also required to assist the designer in producing the structural definition *per se*. It is possible to conceive of a large-scale CASD system which, even at this stage of the ship design, would not need to resort to finite element analysis. Given that FEMs have been used until recently as largely analytical checks of the overall structure, previously sized by traditional formulae, as described in section 6.6, an approach of accessing powerful 'stand alone' analysis suites such as SESAM could be adopted. This approach would have the advantage that a tried finite element analysis system could be used without the need to interface it with the CASD system and its specific hull definition. However, this would firmly relegate the involvement of FEM in ship structural design to an analytical tool brought in after the structural synthesis.

There is a price to pay for the incorporation of a finite element component within a CASD system. First the structural design has to be seen as just a component of the overall design, and this intimately linked into the overall design. This is somewhat inimical to the general trend in technology where there is increased specialisation within structural design and beyond that in finite element analysis applied to marine structures. However, in the wider use of FEM there is an awareness that the major problems in FEM lie in the organisation of the overall design problem and the need to integrate FEM into structural *design* and the total design task.[32] Thus, for example, in the GODDESS system[8] the database, describing a particular ship design, consists of numerical (e.g. weight, space and cost) and geometric (e.g. planes, curved surfaces, curves and the topology of spaces) elements with which the design, audit and analysis programs interact. The structural design features of the GODDESS system are in three parts:

(1) calculation of loading, both quasi-static (BALANC program) and dynamically derived (RAOBLD and MOTION programs);
(2) synthesis of structural configuration, analysis and weight estimation (STRUCT program for defining longitudinal structure and transverse bulkheads and analysing panels of grillages);
(3) three-dimensional finite element analysis described below.

The three components of the structural design capability are linked to

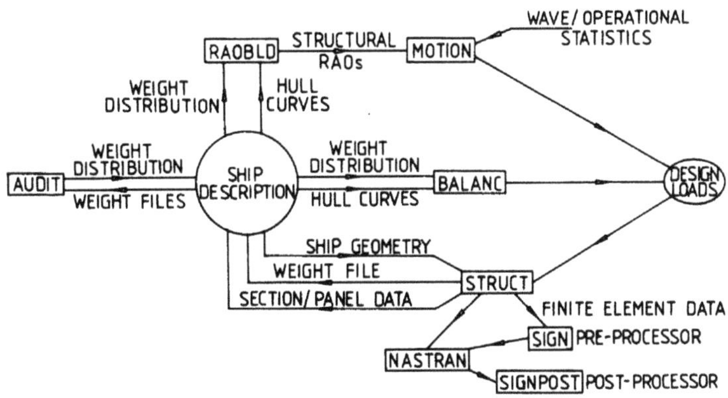

FIG. 6.19. GODDESS: structures programs linked to ship description.[8]

the ship description in the manner indicated by Fig. 6.19, which shows clearly how the finite element feature is an integral part of the overall ship design facility.

Leaving aside the structural loading, synthesis and analysis features of GODDESS described in refs 8 and 29, the finite element analysis capability will be considered in more detail. The program STRUCT, used to define the structure, can also be used to create three-dimensional finite element data automatically from the structural definition. As Fig. 6.19 shows, this data can be directly analysed by MSC NASTRAN or go through a pre-processor (SIGN), a small analysis package (STAMINA), and a post-processor (SIGNPOST), the latter programs being a package produced by Lloyds Register of Shipping. The approach described is not intended to produce 'whole ship' large-scale finite element analysis of the type described in section 6.6, for the reasons previously mentioned. Rather it is intended for analysis of specific structural features during the evolution of the structure. Typically, these might be:

(a) transverse strength between bulkheads to size transverse framing;
(b) superstructure efficiency;
(c) specific structural discontinuities such as deck openings.

This lack of a whole ship capability, along with the need to interface with the specific ship description fundamental to the particular CASD

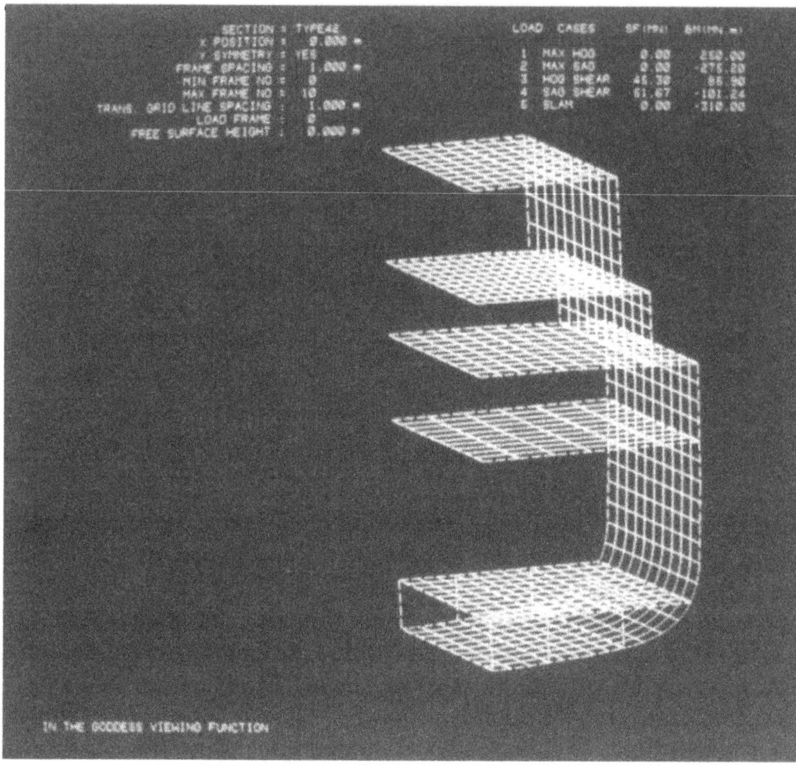

FIG. 6.20. GODDESS: three-dimensional finite element definition of a compartment.[8]

system, could be seen as additional restrictions but they seem more than justified by the enhancement to the structural design facility of the ready access to a finite element analysis capability.

The program STRUCT is used to create finite element data files. The designer can be presented with a three-dimensional display such as Fig. 6.20, which shows a compartment between bulkheads, symmetry having been invoked. This example of one quarter of a frigate compartment has approximately 2000 degrees of freedom when discretised. This is achieved by specifying mesh spacing, element type, loading and boundary constraints on a panel basis to NASTRAN format for subsequent analysis. The loading can be applied both to individual panels, for local loads, and to the overall compartment model, as would be required for primary loads and pressure loading. The capability to undertake finite element analysis of detailed configurations, while they are still evolving, gives the structural designer the ability to undertake meaningful trade-offs between structural considerations and other demands on the design as they arise, such as choice of location of deck openings. Options exploring cost advantages can also be considered, and thus specific optimisation features have not been adopted. It is preferred to provide the designer with the ability to rapidly explore different structure solutions, unconstrained by specific predetermined measures of structural merit.

The approach produced by Hughes *et al.*[5] (SHIPOPT) is in marked contrast to the GODDESS approach in that it is specifically a ship structural design program and is designed to solve large-scale, non-linear, highly constrained, structural optimisations. The cornerstone of SHIPOPT is its special 'design-oriented' FEM, which uses large specially developed elements, each of which corresponds to a principal structural member in a ship such as a stiffened panel or a keel girder. Instead of using the general-purpose elements in more general programs, such as NASTRAN or SESAM, these special elements reduce the complexity of the model. For example a 100 000 ton segregated ballast tanker has been modelled with just 168 design variables and 1080 constraints, of which approximately half are non-linear. Assumptions about structural interactions and boundary conditions can be avoided by analysis of a large portion of the ship rather than a single compartment between bulkheads typically adopted in the GODDESS grillage synthesis previously described. The system has four types of elements:

(a) Stiffened panel elements which can represent large stiffened panels with a single element. Additionally, the stiffness matrix is expressed analytically which speeds evaluation compared with normal numerically integrated elements.
(b) Composite beam elements for modelling bracketed beams by incorporating an effective breadth of plating acting with the beam.
(c) Strut elements for representing pillars and portions of structure which, by in-plane rigidity, contribute to overall stiffness.
(d) Double bottom elements which are modelled as equivalent single plated grillages with wide flange beams and girders, including torsional stiffness incorporated in the stiffness matrix.

The finite element analysis is usually performed in two stages:

(1) a three-dimensional 'coarse-mesh' analysis of the hull, possible due to the sophisticated elements;
(2) a series of two-dimensional 'fine-mesh' analyses of selected areas.

Despite the coarseness of the mesh, all of the important stress values in all of the principal structural members of the ship are obtained. The system is intended to go beyond simply a stress representation by giving a complete limit state analysis. This analysis is hierarchical in that overall adequacy is first assessed and the problem strakes identified. For each strake round the hull girder the load case with the lowest safety margin for a range of modes of failure is given, while beyond that each individual principal member can be similarly interrogated. Thus, the system can be used to obtain rapid assessment of the overall adequacy of a design in its preliminary state, without recourse to extensive detailed large-scale finite element analysis. The approach can also be used to provide the required boundary conditions for regions of structure which the overall analysis indicates as requiring a two-dimensional fine mesh analysis. There are restrictions both in that elements have to be longitudinally prismatic and in that neither stress concentrations nor local details can be dealt with. In this latter respect, the system is less attractive than the approach adopted for the GODDESS system. On the other hand the extension through a reliability analysis, using partial safety factors, to a multiconstrained, non-linear optimisation gives the designer a means of assessing alternative structural solutions.

The two systems outlined above show separate and distinct approaches to using the capability of finite element techniques to assist in the early evolution of a ship's structure. Both exploit modern CAD developments to incorporate the finite element analysis in an interactive manner in the design. The GODDESS system does this within a comprehensive total ship design approach, seeing structural design as an integral component of the overall design, as suggested by the initial sections of this chapter. In contrast, the SHIPOPT approach is specific to the ship structural design problem, and the integration of the finite element analysis is within the full programme of the rational approach to ship structure summarised by Fig. 6.1. It remains to be seen how each will develop; however, it is clear that, either way, the FEM is outgrowing its specifically analytical description and becoming a lynch-pin for the design of ship structures.

6.9. FUTURE USE OF FEM FOR SHIP STRUCTURES

To conclude this chapter on the use of FEM in the design of ship structures, it is sensible to consider briefly the innovations that are likely in ship structural design, and how finite element analysis will contribute to the structural designer's approach to tackling these innovations. Innovations are likely to occur not just in ships but also in the material of which they are constructed and in the knowledge of the structure that is to be designed. Despite all these likely innovations it is still possible to make some specific conclusions on the application of FEM to ship structural design with which to end the chapter.

Future ships will require the structural designer to be ever more inventive in providing structural reliability at an ever-diminishing proportion of the ship's cost. In the conventional monohull this is likely to lead, on the one hand, to adoption of more capable materials and, in contrast, to methods of construction that are cheaper and quicker. Thus the structural designer will be faced with materials, structural properties and failure modes different from common structural steels. The few large ships constructed of reinforced concrete have been said to be structurally and economically efficient, while the glass-reinforced (GRP) mine counter measures vessels have evolved from sophisticated analysis techniques. In both cases, finite element analysis has already had to be employed to solve novel design problems. Beyond the conventional ship, proposals to build ever-larger SWATH vessels have led to extensive finite element analysis of a structural configuration in which it is imperative that a highly efficient structural design is adopted to avoid the ever-increasing displacement resulting from a high structural weight fraction. With the non-displacement lift vessels, hydrofoils and hovercraft, as these increase in size the necessity to keep the structure relatively light becomes more challenging. In all these advanced marine vehicles considerable research is required. This reached an extreme with the US Navy 3KSES (3000 ton Surface Effect Ship), which was a considerable extrapolation in design from current hovercraft/SES technology. Large-scale finite element analyses were undertaken using NASTRAN in both static and dynamic modes for various cushioning loads, while the analysis was backed up with measurements on 100 ton prototypes and SES models in ship tanks.

Coupled with advances in materials and ship types, the knowledge of the behaviour of ship structures will improve. Thus the relatively simple representation of loading adopted in current finite element analysis of ship structures will be increasingly replaced by loadings relevant to the true response of the structure to complex dynamics effects. The current difficulties of integrating the dynamic loading response due to slamming with the random wave action is likely to lead to a need for a dynamically responsive structural model. Following on from the overall response is a need for the finite element model to more accurately represent the behaviour of the thin shell and the stiffeners under this dynamic load. Elements will need to model the many modes of structural response so that the various non-linear effects and aspects such as fracture mechanics can be readily predicted. The SHIPOPT system just outlined attempts to integrate into a finite element analysis a limit state approach which is then extended to a full reliability analysis and a fatigue analysis. This amounts to linking into a finite element method many of the issues raised in the early sections of this chapter. However, the price paid for this is the sophistication of the elements and their restricted applicability to conventional steel structures.

It is also inevitable that general advances in FEM applied to structures will be adopted for ship structures. Already there are facilities being developed that are gradually being introduced into general usage. Automatic data generation, coupled with mesh generation and automatic discretisation of loads, greatly enhances the user-friendliness of finite element systems; particularly as graphic display of the resultant model becomes readily available. Improved element features are being developed, such as membrane elements and plate and shell elements having nodes of varying degrees of freedom. Additionally, there is a need for better prediction of accuracy and of stability of solutions. Systems already exist which provide vector displays for post-processors and also deformed shape displays which enable the user to rapidly assess the areas of prime concern instead of ploughing through vast quantities of output, a task that was all too familiar with the early large-scale ship structural analyses such as that for the *Invincible* design.[24] The interfacing of finite element systems with CASD has already been outlined, and it is clear from the rate of development of CAD that commensurate improvements will occur in the FEM/CAD applied to ship structural design. It has already been mentioned that the task of integrating FEM into structural and overall design is the single major challenge for finite element system designers and that the problem is not just technical but also organisational and psychological.[33] Progress is not likely to be simple and straightforward in achieving the highly capable user-orientated finite element design systems of the future.

In conclusion, while finite element techniques applied to ship structures have changed the way in which ship structures are analysed and, increasingly, the manner in which they are designed, out of all recognition, limitations remain:

(a) the picture conveyed from analysis is usually in terms of field stresses, whereas failure may arise from local detailing;
(b) the analysis of whole ship structures is complex and expensive, and so it is rarely adopted unless the design is novel;
(c) collapse mechanisms are likely to be non-linear and thus require expensive modelling and computation;
(d) the finite element approach is essentially analytical and, despite CAD, not design-oriented;
(e) at a price, any structural problem can be contemplated by the finite element analyst,[33] but
  (i) any novel problem must be tackled experimentally as well as analytically, and
  (ii) the solution is largely dependent on the skill of the engineer in preparing the idealisation, and this involves an intimate understanding, foremostly of the nature of the structure he is designing, the way it responds to the loading, and the concept of structural safety his design must meet.

REFERENCES

1. CALDWELL, J., Seaborne structures, in *Engineering Structures,* eds P. S. Bulson, J. B. Caldwell and R. T. Severn, University of Bristol Press, 1983.
2. LLOYDS REGISTER OF SHIPPING, *Rules and Regulations for the Construction of Steel Ships,* London, 1987.
3. CALDWELL, J. B., *Structures and Materials; Progress and Prospects,* Royal Society, London, 1972.
4. International Ship Structures Congress, *Report of Committee VI,* 1976.
5. HUGHES, O. F., JANAVA, R. T. and WOOD, W. A., SHIPOPT—A CAD system for rationally based ship structural design and optimisation, in *Computer Applications in the Automation of Shipyard Operation and Ship Design IV, ICCAS 1982,* Rogers, D. F., Nehrling, B. C. and Kuo, C. (Eds), North Holland, Amsterdam, 1982.
6. ANDREWS, D. J., An integrated approach to ship synthesis, *Trans. R. Inst. Nav. Archit.,* 1985.

7. ANDREWS, D. J. and BROWN, D. K., Cheap warships are not simple, *Symposium on Ship Costs and Energy,* New York, Sept. 1982, Soc. Nav. Archit. Mar. Engrs, 1983.
8. PATTISON, D. R., SPENCER, R. E. and VAN GRIETHUYSEN, W. J., The computer aided ship design sytem GODDESS and its application to the structural design of Royal Navy warships, Rogers, D. F., Nehrling, B. C. and Kuo, C. (Eds), ICCAS, Annapolis, 1982.
9. SALVESEN, N., TUCK, E. O. and FALTINSEN, O., Ship motions and sea loads, *Trans. Soc. Nav. Archit. Mar. Engrs,* 1970.
10. JENSEN, J. J. and PEDERSEN, P. T., Wave-induced bending moments in ships—a quadratic theory, *Trans. R. Inst. Nav. Archit.,* 1978.
11. FAULKNER, D., Semi-probabilistic approach to the design of marine structures, *Extreme Loads Response Symposium,* Arlington, Soc. Nav. Archit. Mar. Engrs, 1981.
12. KENDRICK, S., The structural design of supertankers, *Trans. R. Inst. Nav. Archit.,* 1970.
13. CALDWELL, J. B., Ultimate longitudinal strength, *Trans. R. Inst. Nav. Archit.,* 1965.
14. SMITH, C. S., Compressive strength of welded steel ship grillages, *Trans. R. Inst. Nav. Archit.,* 1975.
15. MANSOUR, A. E. and FAULKNER, D., On applying the statistical approach to extreme sea loads and ship hull strength, *Trans. R. Inst. Nav. Archit.,* 1973.
16. FAULKNER, D., Discusson on ref. 14.
17. International Ship Structures Congress, *Report of Committee VI,* 8th Meeting, Paris, 1982.
18. BLOCKLEY, D. I., *The Nature of Structural Design and Safety,* Ellis Horwood, Chichester, 1980.
19. YUILLE, I. M. and WILSON, L. B., Transverse strength of steel hulled ships, *Trans. R. Inst. Nav. Archit.,* 1960.
20. PAULLING, J. R. and PAYER, M. G., Hull–deckhouse interaction by finite element calculations, *Soc. Nav. Archit. Eng.,* 1968.
21. SMITH, G. K. and WOODHEAD, R. G., A design scheme for ship structures, *Trans. R. Inst. Nav. Archit.,* 1973.
22. MCCALLUM, J., The strength of fast cargo ships, *Trans. R. Inst. Nav. Archit.,* 1975.
23. ARALDSEN, P. O., Examples of large scale structural analysis—ESSO Norway, in *Application of Computerised Methods in Analysis and Design of Ship Structures, Marine Structures and Machinery,* Røren, E. M. (Ed.), Det Norske Veritas, Oslo, 1972.
24. HONNOR, A. F. and ANDREWS, D. J., *HMS Invincible*—first of a new genus of aircraft carrying ships, *Trans. R. Inst. Nav. Archit.,* 1981.
25. MCVEE, J. D., Finite element study of hull–deckhouse interaction, *Comput. Struct.,* **12**(4), 1980, October.
26. AL-SANI, M., The application of the finite element techniques to the transverse strength of ships, Ph.D. Thesis, University of London, 1980.
27. SMITH, C. S. and KIRKWOOD, W., Influence of initial deformations and

residual stresses, in *Steel Plated Structures*, Dowling, P. J., Harding, J. E. and Frieze, P. A. (Eds), Crosby Lockwood Staples, London, 1977.
28. Jackson, R. I. and Frieze, P. A., Design of deck structures under wheel loads, *Trans. R. Inst. Nav. Archit.*, 1981.
29. Holmes, S. J., The application and development of computer systems for warship design, *Trans. R. Inst. Nav. Archit.*, 1981.
30. Nethercote, W. C. E. and Schmitke, R. J., A concept exploration model for SWATH ships, *Trans. R. Inst. Nav. Archit.*, 1986.
31. Chalmers, D. W., Preliminary structural design of warships, *Trans. R. Inst. Nav. Archit.*, 1986.
32. Fenves, S. J., Future directions of structural engineering applications, *Comput. Struct.*, **10**(1–2), 1979.
33. Clough, R. W., Finite element method after twenty five years: a personal view, *Comput. Struct.*, **12**(4), 1980, October.

Chapter 7

Finite Element Analysis and Design of Thin-walled Structures in the Offshore Industry

NAHID JAVADI
Wimpey Offshore, London, UK

7.1. INTRODUCTION

The use of thin-walled steel structures in the offshore industry has been extensively developed in the North Sea over the past 20 years.

Advanced analysis techniques, materials and manufacturing requirements have been developed and integrated to overcome the more severe environmental conditions and increasing water depth requirements of recent years. The most major advanced analysis technique, the finite element method, has been widely used throughout the design process for the effective and accurate appraisal of the key highly stressed areas. Thin-shell and plate finite elements are the most frequently used finite elements in the estimation of the stress distribution in the critical sections of offshore structures.

The next section of this chapter presents a review together with illustrations of structural details that are most frequently analysed, using thin-shell finite elements. The following section describes the theoretical background to thin-shell finite elements. A further section discusses the loading and boundary conditions, while the final section reviews the pre- and post-processing finite element facilities.

7.2. SOME CATEGORIES IN THE APPLICATION OF THIN-SHELL FINITE ELEMENTS IN THE OFFSHORE INDUSTRY

With the rapidly reducing computer costs over the past decade and the increasing availability of software, not only for analysis but also for data preparation, data checking, post-processing and load combination facilities, there is now an increased incentive to perform a greater number of finite element analyses in an offshore design office. Many localised areas of steel jackets, self floaters, subsea templates, risers, integrated decks and topside structures are now even more thoroughly investigated using thin-shell finite elements.

Finite element analyses are usually performed on offshore structures in routine stress analysis, load path investigation and ultimate strength design, each of which will be looked at separately.

7.2.1. Routine Stress Analysis

The most commonly used offshore structures are the fixed platforms, with their substructures built up as a space frame of tubulars and then piled through the sea bed, as shown in Fig. 7.1.

In fixed platform design, the joints between tubular members are critical for the platform's fatigue life. Due to the complex geometry of joints, prediction of the stress concentration factors is extremely difficult and could not be accurately appraised without a finite element analysis. Increasingly, the analysis of complex multibraced tubular connections is carried out to confirm design assumptions used in the fatigue analysis and as necessary, to model design changes enabling an optimum arrangement to be achieved by the engineer.

In Fig. 7.2, a finite element model of a leg node is illustrated. The six ring stiffeners and the four incoming braces have been modelled using thin-shell finite elements. The finite element model was generated using the interactive graphic's program PATRAN-G.[1] Refinement of the mesh was applied, where necessary, to accurately assess the high stress gradients near the intersection of the braces and the ring stiffeners with the chord.

In general, elements immediately adjacent to brace/chord intersection lines are sized such that the element edges are located as near as practicable to the toe of the weld. The principal stresses are the computer output at the inside and outside shell surfaces and are used for predicting hot-spot stress at the weld toe. The finite element

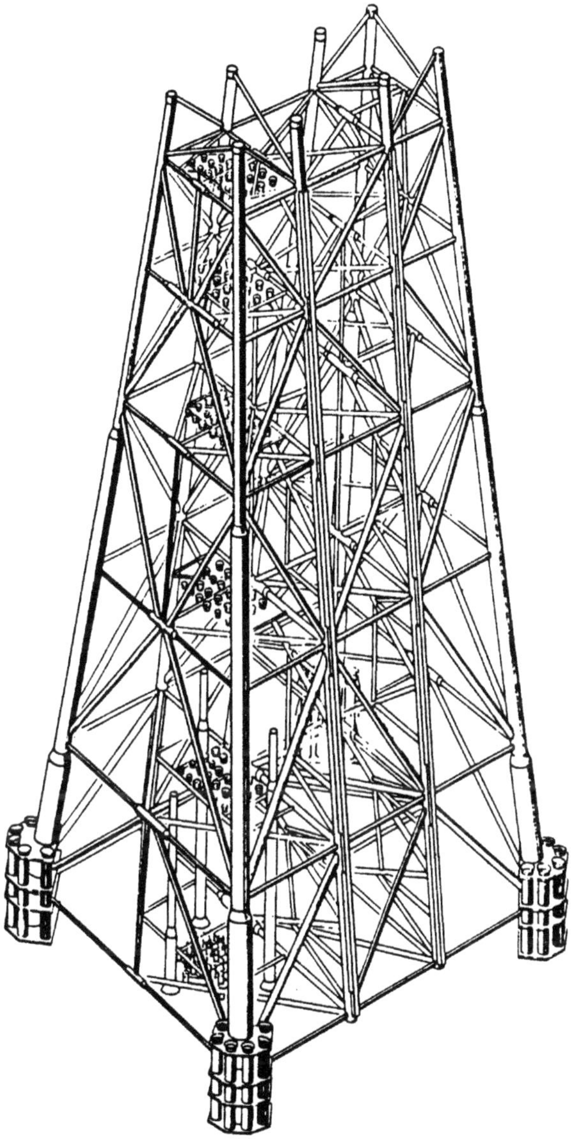

FIG. 7.1. Fixed offshore structure.

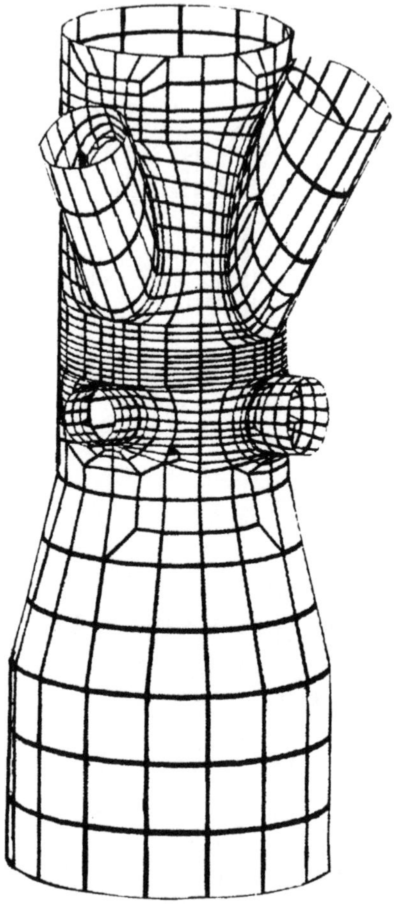

FIG. 7.2. Finite element model of a leg node.

analysis was performed with the program ASAS-H.[2] Thin-shell surfaces are modelled using the Semi Loof shell element GCS8 developed by Irons.[3]

The stress concentration factors for all positions are plotted against the angle around the intersection lines, as shown in Fig. 7.3, using an in-house post-processing routine.

Stiffened plated structures, of the type that is common to the steel legs of hybrid platforms or floater pontoons and primary columns of tension leg platforms, are very often subject to local fractures in the

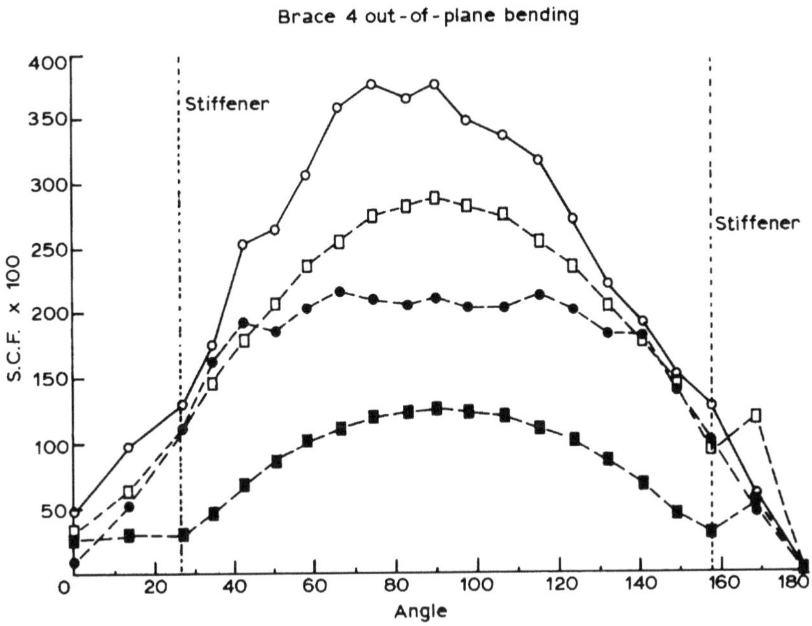

FIG. 7.3. Stress concentration versus angle. ○, brace (unstiffened). □, chord (unstiffened). ●, brace (stiffened). ■, chord (stiffened).

weld areas. Periodic inspection of these structures usually reveals fatigue cracks at the toe of the welds in the hot-spot regions. In order to study the growth of these cracks using the fracture mechanics principle, finite element analysis is employed to identify the stress distribution in these areas. The finite element model of a stiffened plated/cylindrical steel leg of a hybrid structure is illustrated in Fig. 7.4. Thin-shell elements are used to model the plated sections, and doubly curved beam elements are used to model the stiffeners. The incoming members are not represented in this global model, but the load transferred by the incoming members is input to the appropriate interaction points.

Localised finite element meshes of the four corners are later generated using thin-shell elements to allow a more refined representation of internal stiffening and incoming members, as shown in Fig. 7.5. Plots of the stress distributions for the zones of interest are generated using PATRAN-G post-processing facilities. The stress distribution of a top cover plate can be seen in Fig. 7.6.

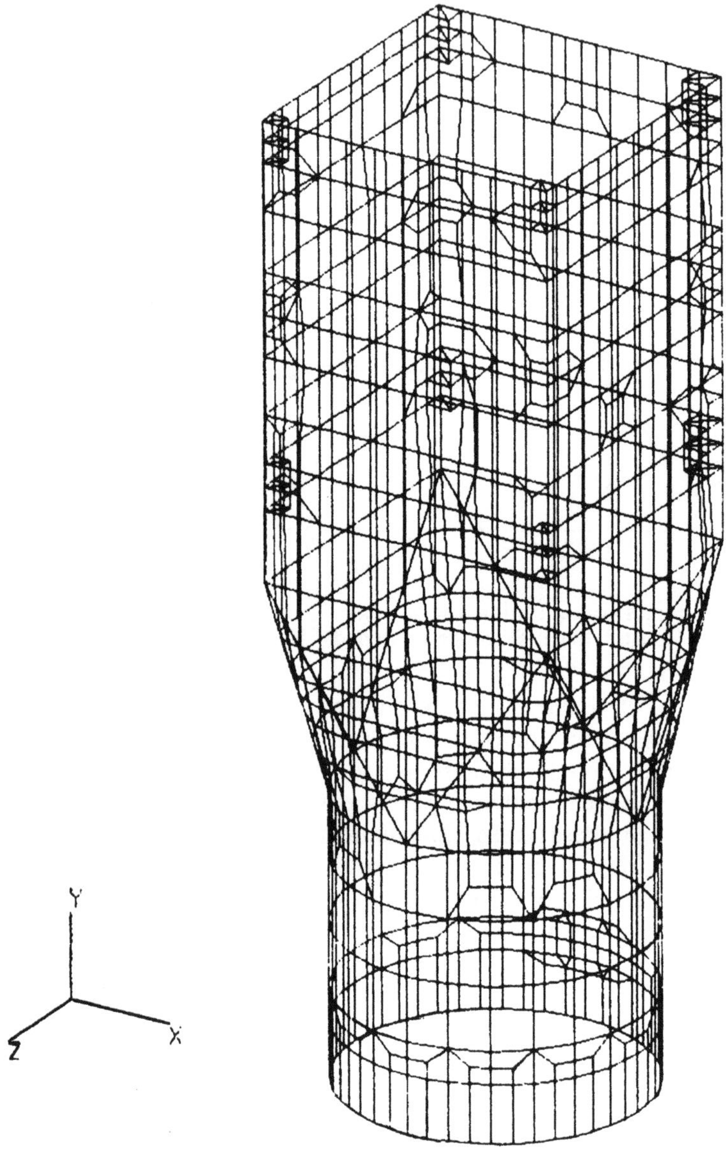

FIG. 7.4. Finite element model of a stiffened leg.

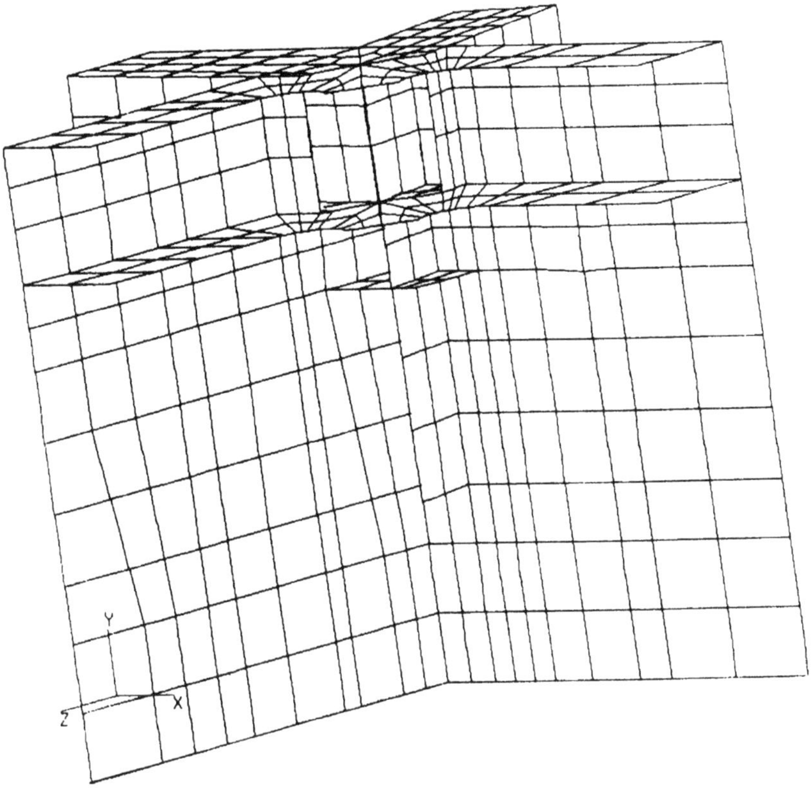

FIG. 7.5. Finite element model of a corner detail.

7.2.2. Load Path Investigation

The proportions and complexity of a structural detail often exceed the limits of existing design practices. In these circumstances, finite element methods are used to establish a thorough understanding of load flow behaviour in the required section.

One of the most interesting examples in this category is an investigation of load paths in pile/sleeve details. The design of this structural detail is complicated by the large forces that have to be transferred through a system of vertical and horizontal plates into the piles. Therefore, the lower bottle section of a leg and six sleeves are modelled fully, as represented in Fig. 7.7. The leg shell, upper and

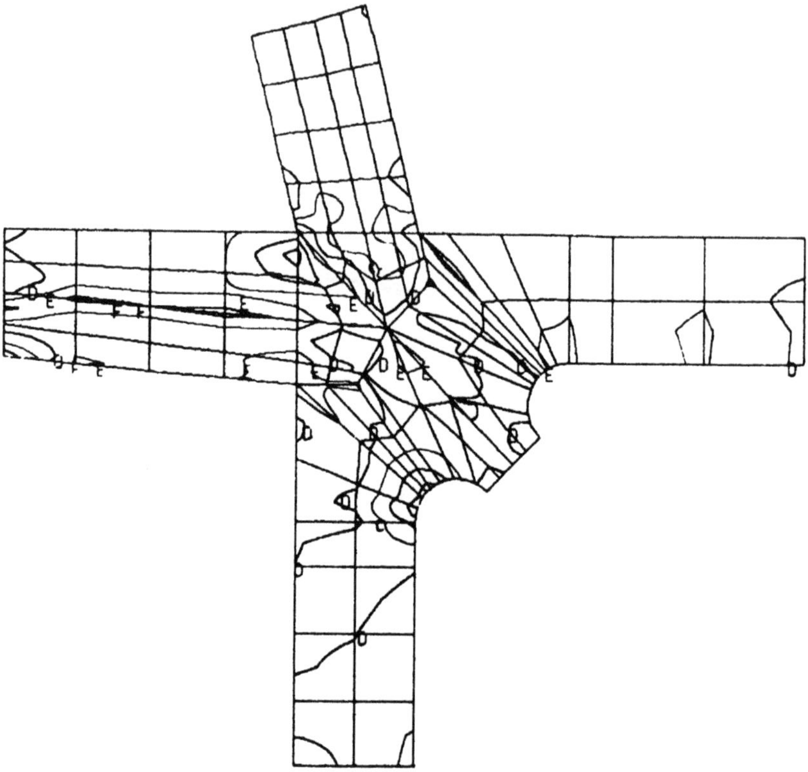

FIG. 7.6. Stress contouring of top cover plate.
Top surface principal stress P1

48·9 = A
36·0 = B
23·2 = C
10·4 = D
−2·46 = E
−15·3 = F
−28·1 = G

FIG. 7.7. Finite element model of pile/sleeve detail.

lower vertical shear plates, pile sleeve skirts and incoming braces are modelled using the quadrilateral shell elements GCS8 and the triangular shell elements GCS6. The full model takes no advantage of symmetry and therefore suppressions are simply applied at the boundary and at the ends of the axial and lateral springs representing the six piles. Load paths are evaluated from the element stress distributions for the applied unit load cases.

In the course of a structural design, it may be necessary to justify the serviceability of a structural section to the certifying agency. This is particularly essential when a new material or design is used. For example, when pad ears were incorporated into the casting of caisson bottles, in order to facilitate their removal after use, a finite element analysis was performed to satisfy the basic structural requirements of strength and serviceability. Because of a large diameter of caisson and low wall thickness to diameter ratio, thin-shell elements GCS8 were used to generate a half model, taking advantage of symmetry, as shown in Fig. 7.8.

7.2.3. Ultimate Strength Design

The integrity and serviceability of offshore structures that may be subjected to accidental loads, such as ship impacts has, during the past few years, been added to the existing responsibilities of the design engineer. As a consequence of this, there is a demand for analysis capabilities which are able to follow the structural behaviour up to the structure's ultimate load-carrying capacity. For this, non-linear programs are specially developed incorporating the analysis of shell structures, including geometric and material non-linearities. A non-linear version of a Semi Loof shell element is implemented in the LUSAS program.[4] This element has been extensively extended to include both material and geometric non-linearities. The plasticity effect is taken into account by a rigorous multilayer formulation, using five layers to represent the shell thickness. The geometric non-linearity is taken into account by using a total Lagrangian formulation. The solutions of the non-linear equilibrium equations are carried out iteratively using load or displacement increments and Newton–Raphson, or modified Newton–Raphson iterations.

The non-linear capability of this programme can be demonstrated in the analysis of a T-joint. A quarter model is generated utilising 49 non-linear Semi Loof shell elements, and is loaded by axial point

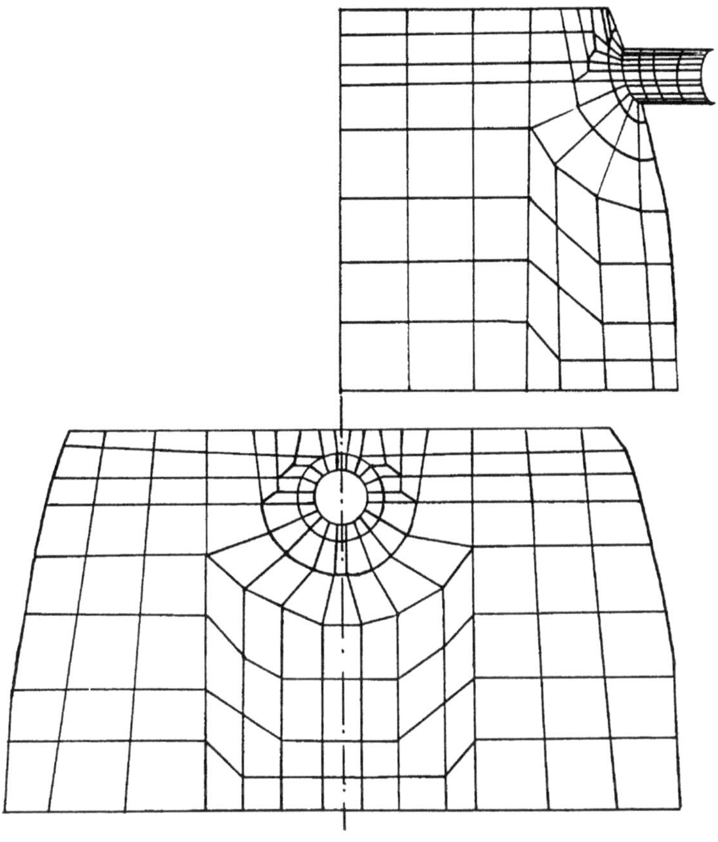

FIG. 7.8. Finite element model of caissons incorporating padears.

loads. The deformed geometry of the brace and the chord wall are shown in Fig. 7.9.

Cylindrical shells of high diameter to wall thickness ratio are used in deepwater fixed platforms and are subject to local buckling. Because of the very high hydrostatic pressure and high axial compressive loads acting on deepwater fixed platforms, it is essential to validate the current state of the art in the design of the hydrostatic collapse of tubular members.

A local buckling analysis of a single bay stiffener of a leg of a deepwater fixed platform, subjected to hydrostatic pressure and axial

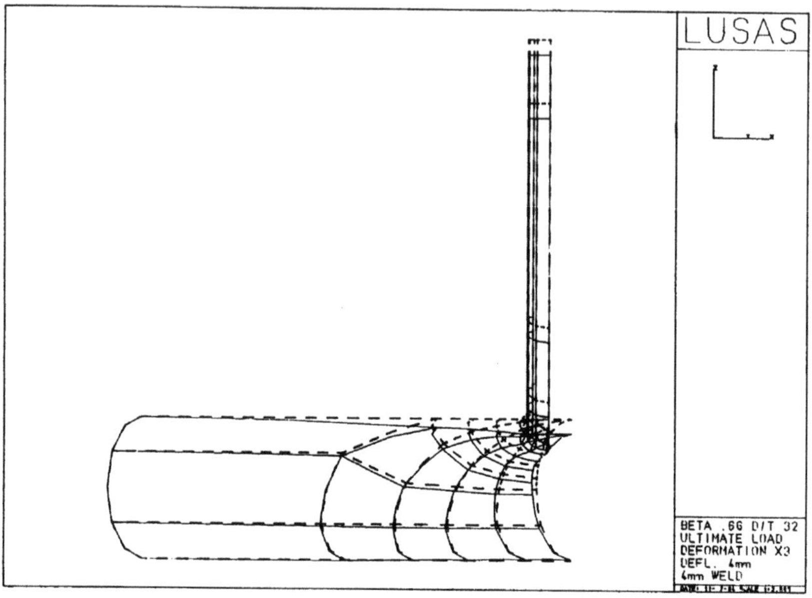

FIG. 7.9. Ultimate load analysis of T-joint.

compressive loads, can be performed using non-linear shell elements. Non-linear beam elements, which can model initial imperfections and residual stresses, are used to model the ring stiffeners.

It was found that the finite element analysis was capable of considering different parameters that affect the strength of the leg, namely fabrication tolerances and residual stresses. The buckling mode of failure, inter-stiffener buckling and stiffener buckling were accurately predicted and were validated against available experimental data.

7.3. THEORETICAL BACKGROUND TO THIN-SHELL ELEMENTS

The majority of general-purpose finite element programs provide the user with a wide range of elements for the solution of thin-shell structures. Four alternative forms of finite element representation can be identified from the multiplicity of shell elements available. These are:

(a) the faceted form using flat elements;
(b) curved shell elements formulated from appropriate thin-shell theories;
(c) isoparametric solid elements specialised to tackle thin shells by applying, in discrete form, approximate thin-shell assumptions (for example, Kirchhoff's normality hypothesis);
(d) specialised elements to deal with axisymmetric shells with either axisymmetric or asymmetric loadings.

A very crude and often extremely inaccurate way of analysing a shell is to approximate the surface by the discrete flat elements (category (a)). The physical approximation introduced by using flat elements leads to decoupling of the membrane action and bending action within each element. For thin-shell surfaces, rapid stress gradients invariably occur and it is necessary to employ very fine subdivisions in order to obtain the requisite accuracy.

Some quadrilateral and triangular finite elements based on thin-shell theory (category (b) above) exist, and use corner and mid-side nodes with cubic or even quintic displacement components. This usually makes the elements 'over stiff' and introduces the curvature freedoms* continuity of bending strains across element interfaces. If the element interface occurs at a change in thickness, this will lead to incorrect discontinuity in the moment terms. Engineers usually find these elements far too complicated to use for most of the shell problems.

More recent elements, those in the category (c), are generally curved triangles and curved quadrilaterals with nodes at the corners, the midsides, along the four edges and internally (Loof points). They are essentially degenerate isoparametric solid elements. An important feature of these elements is that the rotations are interpolated independently of the translation at the Loof nodes, by constraining the rotations along the sides at each of the Loof nodes, or by imposing the Kirchhoff hypothesis in the form of zero shears at the Loof nodes.

The degrees of freedom are the three Cartesian translations X, Y and Z at each of the corner and midside nodes, plus the normal rotations R_1 and R_2 at the two Loof nodes on each side of the element. For identification purposes, the two rotations are treated as mid-side

* The ability of any node to translate and rotate about X, Y, Z axes is represented by six degrees of freedom including three translational Dof(X, Y, Z) and three rotational Dof (about X, Y, Z axes).

ASAS USER MANUAL (GCS8)

Generally Curved Quadrilateral Thin Shell Element with Varying Thickness Capable of Modelling Discontinuities in Curvature and Thickness

Number of nodes 8 (4 corner, 4 mid-side)

Nodal co-ordinates x, y, z (may be omitted for mid-side nodes on straight edges). Each mid-side node has a tolerance of side-length/10 about the true mid point.

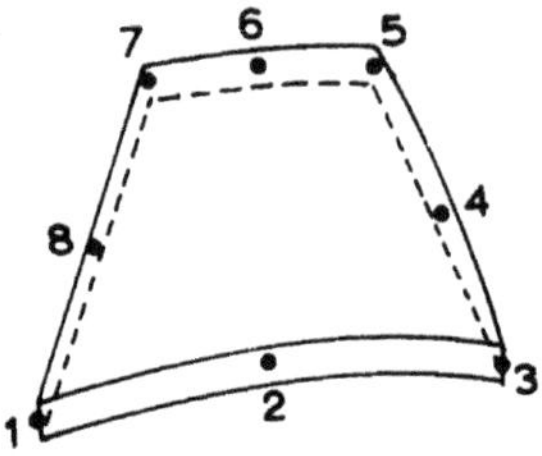

Degrees of freedom X, Y, Z at corner nodes. X, Y, Z, R_1, R_2 at mid-side nodes. (R_1 and R_2 are rotations about the edge at the 'Loof' points, and provide the bending compatibility between elements. R_1 is between the lower-numbered corner node and mid-side node, R_2 between the mid-side node and higher-numbered corner node. The $+ve$ direction of R_1 and R_2 is clockwise looking from the lower-numbered node to the higher-numbered node.)

Geometric properties

t_1 Thickness at node 1 (>0·0)
t_2 Thickness at node 2
t_3 Thickness at node 3
t_4 Thickness at node 4
t_5 Thickness at node 5
t_6 Thickness at node 6
t_7 Thickness at node 7
t_8 Thickness at node 8

(t_2 to t_8) t_2 to t_8 may be omitted for an element with uniform thickness t_1; the second Geometric Properties card must then also be omitted.

Material properties

isotropic:

E Modulus of elasticity.
ν Poisson's ratio.
α Linear coefficient of expansion.
ρ Density (mass/unit volume); see Appendix B.
(α and ρ are not always needed.)

anisotropic:

ρ Density (mass/unit volume); see Appendix B.
21 Coefficients of the local three-dimensional stress–strain matrix.
3 Linear coefficients of expansion $\alpha_{x'x'}$, $\alpha_{y'y'}$, $\alpha_{x'y'}$ related to the local axes.

Load types Standard types listed at the start of this Appendix.
Pressure loads ($+ve$ for $+ve$ local Z' direction).
Temperature loads.
Face temperatures (Face 1 is on the $-ve$ local Z' side).
Centrifugal loads.
Distributed load Types ML1 + ML2.

FIG. 7.10. Semi Loof shell element.

Mass modelling Consistent mass (used by default), lumped Mass.

Stress output Membrane stresses $\sigma_{x'x'}$, $\sigma_{y'y'}$, $\sigma_{x'y'}$ and bending moments/unit length $M_{x'x'}$, $M_{y'y'}$, $M_{x'y'}$ are related to local axes.

Node numbering The nodes are listed in cyclic order, clockwise or anti-clockwise, starting at a corner node.

Anisotropic matrix

$$\begin{bmatrix} \sigma_{x'x'} \\ \sigma_{y'y'} \\ \sigma_{x'y'} \\ M_{x'x'} \\ M_{y'y'} \\ M_{x'y'} \end{bmatrix} = \begin{bmatrix} C_1 & C_2 & C_4 & C_7 & C_{11} & C_{16} \\ \cdot & C_3 & C_5 & C_8 & C_{12} & C_{17} \\ \cdot & \cdot & C_6 & C_9 & C_{13} & C_{18} \\ \cdot & \cdot & \cdot & C_{10}t^3 & C_{14}t^3 & C_{19}t^3 \\ \cdot & \cdot & \cdot & \cdot & C_{15}t^3 & C_{20}t^3 \\ \cdot & \cdot & \cdot & \cdot & \cdot & C_{21}t^3 \end{bmatrix} \begin{bmatrix} \sigma_{x'x'} \\ \sigma_{y'y'} \\ \sigma_{x'y'} \\ W_{x'x'} \\ W_{y'y'} \\ W_{x'y'} \end{bmatrix}$$

The coefficients C_7, C_8, C_9, C_{11}, C_{12}, C_{13}, C_{16}, C_{17} and C_{18} are always zero. Note that the coefficients C_{10}, C_{14}, C_{15}, C_{19}, C_{20} and C_{21} do not contain the thickness term.

Local axes Local X' is a curvilinear line on the shell surface. At any point it is defined by the intersection of the shell with a plane containing the surface normal and a line parallel in space to the straight line from node 1 towards node 3. Local Y' lies in the shell surface, $+ve$ towards node 6. Local Z' forms a right-handed set with local X' and local Y'.

Sign conventions Direct stresses $\sigma_{x'x'}$, $\sigma_{y'y'}$, $\sigma_{x'y'}$ $+ve$ as shown.
Bending moments $M_{x'x'}$, $M_{y'y'}$, $M_{x'y'}$ $+ve$ as shown.

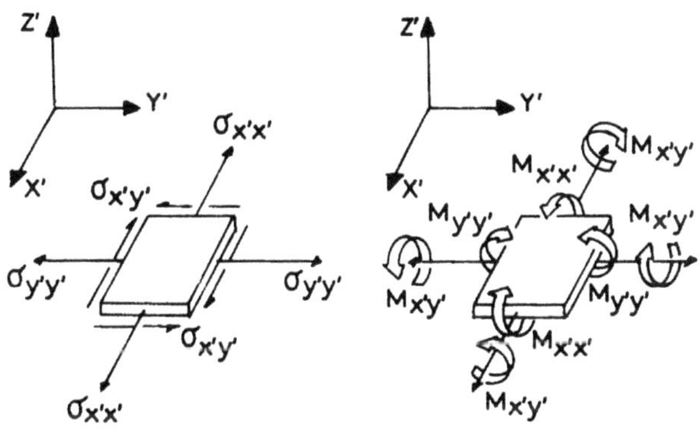

For data examples, see ref. 6.

FIG. 7.10—*contd.*

node variables, effectively giving five degrees of freedom at the mid-side nodes. The geometric input consists of the Cartesian coordinates of the corner and the mid-side nodes. The Semi Loof shell elements are easy to use for practical problems and can be used together with membrane elements and compatible curved beam elements. These elements are implemented in the ASAS element library as GCS8 (eight-noded quadrilateral) and GCS6 (six-noded triangle), as shown in Fig. 7.10.

For axisymmetric shell analysis, meridionally curved elements are available. The axial displacements are interpolated linearly and there is cubic interpolation of normal displacements. The degrees of freedom consist of two translations and two derivatives, at each of the two ends of the element.

For the majority of structural modelling in the offshore design office, the Semi Loof shell elements are an obvious choice. They are capable of accurately representing sharp corners, intersecting branches and complex geometry with coarser meshes, thus saving time and expense in the analysis. These elements have also been chosen by UKOSRP after careful appraisal of available elements in the ASAS system.

7.4. LOADING AND BOUNDARY CONDITIONS

The accuracy of the finite element analysis is increased if an accurate representation of the loadings is used in the analysis. Loads applied to offshore structures can be classified into one of the following three groups:

(a) structural loadings, typically nodal loads;
(b) applied displacements, where values of the displacements on the model boundaries are given as initial conditions;
(c) initial strain or initial stress loads, such as lack of fit.

Structural loads are often derived in the form of point loads from the global frame analysis, subjected to the weight of the module support frame, the weight of equipment and the environmental loads. Loads in the form of distributed loads, i.e. hydrostatic pressure, wind load, etc., can be approximated by the equivalent nodal loads, using the element shape functions.

Boundary displacements along the edges of a local model are known in cases in which the engineer has previously studied the behaviour of

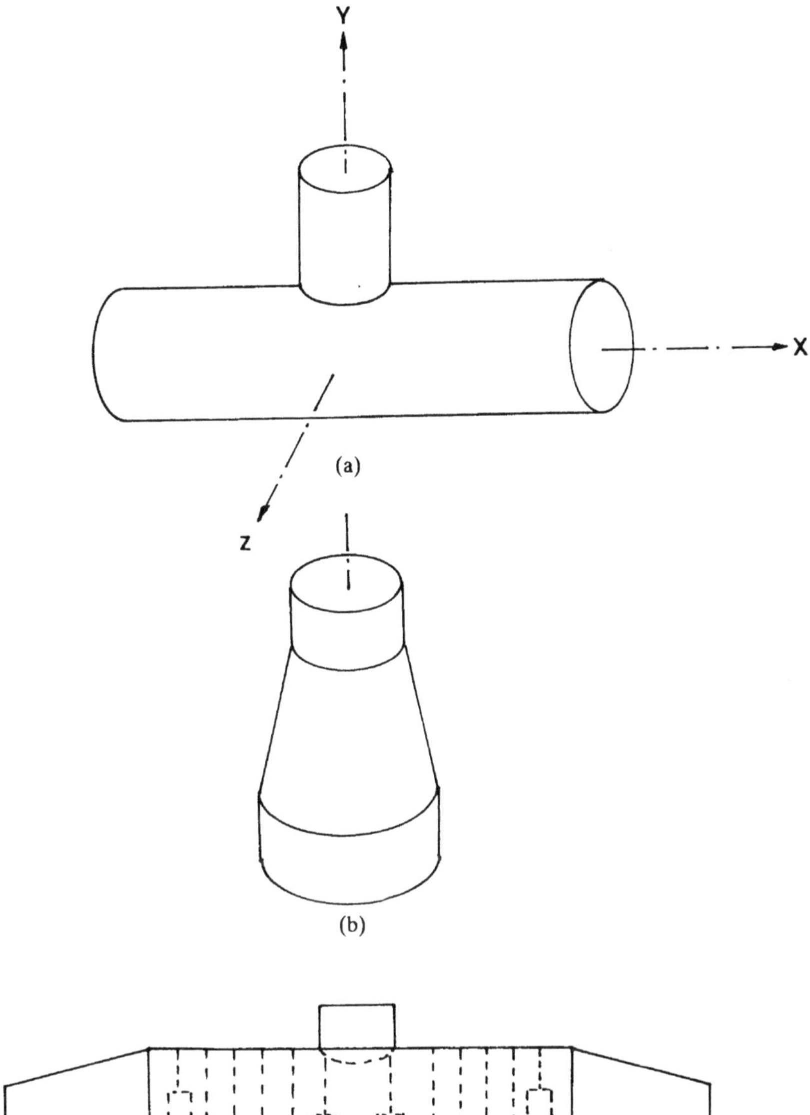

FIG. 7.11. Conditions of symmetry.

relevant global area. Combinations of nodal loads and boundary displacements can be applied by using the principle of superposition.

There are, on many occasions, ways in which the structure's various symmetries can be used to reduce the mesh size and the costs; although the use of planes of symmetry results in additional load cases. There is a net cost saving, and the finite element solution cost is related to the number of nodes making up the problem. Various forms of symmetries can occur, but the main forms are:

(a) Mirror-image symmetries as illustrated in Fig. 7.11(a).
(b) Axisymmetries, where the structure is obtained by rotating it about a central axis, as in Fig. 7.11(b).
(c) Repetitive symmetry, as illustrated in Fig. 7.11(c). In this case the structure is composed of continuously repeated sections, if the end effects are ignored.

The loading on the structure may also have symmetries, and generally loads can be considered as the sum of a series of separate load cases, where each term in the series is either symmetric or antisymmetric. When taking advantage of symmetry, extra care and effort must be employed in the handling of all of the boundary conditions and load cases, to ensure that mistakes do not occur.

7.5. PRE- AND POST-PROCESSING

Significant advances have been made in the past decade in the development and application of pre- and post-processing programs for stress analysis. The advances made in computer technology, hardware and software, provided the foundation from which many general-purpose processing programs have evolved. After a decade of development, a variety of these programs are currently being used in offshore consulting offices. The hardware requirement and graphic's capabilities vary considerably from one processing package to another.

The increasing importance of stress analysis and the sense of urgency in producing accurate results prompts the use of versatile and powerful processing programs, together with colour graphic capabilities. A number of factors which are specifically important are listed in the following sections.

7.5.1. Solid Modelling Capabilities

For stress analysis, the following modelling features facilitate the easy checking of the model geometry and the shapes and sizes of the meshes:

(a) consistent construction options (lines, surfaces and solids);
(b) surface intersection and arbitrary geometry;
(c) topology checking for consistent geometry;
(d) special co-ordinate frames (Cartesian, cylindrical and spherical);
(e) part naming;
(f) blend, break, merge and cursor options, selective identification,

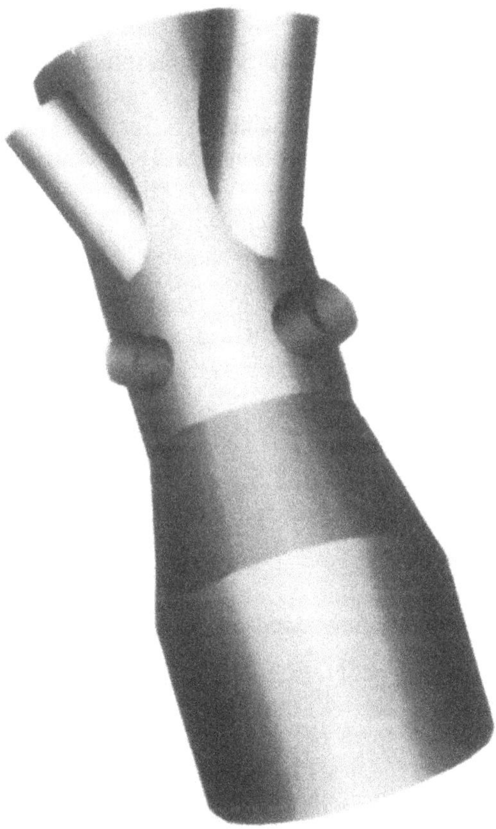

FIG. 7.12. Solid model of a leg node.

erase and delete options, capabilities to translate, rotate, scale, mirror and glide;
(g) efficient hidden-line removal algorithms;
(h) complete viewing control;
(i) orthographic perspective and 'walk through' viewing options.

A typical three-dimensional colour display of a stiffened leg node with four incoming members is presented in Fig. 7.12.

7.5.2. Mesh Generation Capabilities

The following mesh generation and model verification features are essential for the accurate use of finite elements, and the subsequent analysis:

(a) automatic uniform/non-uniform grid element generation and the transition and selective editing of nodes and elements;
(b) automated nodal equivalencing;
(c) node removal, compaction and resequencing element compaction;
(d) element free edge or crack checking;
(e) aspect ratio, warp and interior angles;

The finite element mesh of the leg node, shown in Fig. 7.12, was constructed using thin-shell elements (quadrilateral GCS8 and triangular GCS6), and was shown in Fig. 7.2.

7.5.3. Post-processing Facilities

In addition to the pre-processing options, and after the conclusion of the analysis, the following features are often necessary:

(a) deformed geometry plots, contour plots of stresses and displacement results;
(b) element results colour coded or colour fringe plots.

The post-processing program SHELLST[5] and the graphic capabilities of PATRAN-G can be combined to produce fringe plot of the principal stress factors for a K-joint.

The incidence of error in data input has been greatly reduced, since the application of graphical capabilities. Also, better output presentation has been achieved, since the post-processor can sift the data and produce more focused, attractive and less voluminous output.

7.6. CONCLUSIONS

This chapter has presented an evaluation of thin-shell finite element analyses performed in the offshore engineering design offices. Examples of analyses carried out during the author's involvement with offshore structures are reviewed, highlighting different categories of structural details.

The interactive pre- and post-processing graphic system used provided versatile modeiling, mesh generation, documentation and better output presentation capabilities.

On-going development of computer-aided engineering (CAD/CAM) software systems and its subsequent linkage with the existing finite element package will provide a means of automating all aspects of offshore engineering. This will introduce a considerable degree of automation in the routine finite element analysis of structural details, thus saving time and money, and improving accuracy.

ACKNOWLEDGEMENTS

The author wishes to thank Mr Gail Lewis, Technical Director of John Brown Offshore Structures Ltd, for permission to write this chapter.

REFERENCES

1. *PATRAN-G Manual,* PDA Engineering, 1560 Brookhollow Drive, Santa Ana, CA 92705, USA. 15 March 1984.
2. *ASAS User Manual,* Version H, Issue 3, Atkins Research and Development, Woodcote Grove, Epsom, Surrey KT18 5BW, UK. June 1983.
3. Irons, B. M., The Semi Loof shell element, *Report CIR/124/70,* University College of Wales, Swansea, UK. 1973.
4. *LUSAS User Manual,* FEA Ltd, Forge House, 66 High St, Kingston upon Thames, Surrey KT1 1HN, UK. May 1986 Version 86.07.
5. *SHELLST User Manual,* Atkins Research and Development, Woodcote Grove, Epsom, Surrey KT18 5BW, UK. 1983.
6. Irons, B. M. The Semi Loof shell element. In *Finite Elements for Thin Shells and Curved Members,* John Wiley, New York, 1976.

Index

For Product Safety Concerns and Information please contact our EU representative GPSR@taylorandfrancis.com
Taylor & Francis Verlag GmbH, Kaufingerstraße 24, 80331 München, Germany

www.ingramcontent.com/pod-product-compliance
Ingram Content Group UK Ltd.
Pitfield, Milton Keynes, MK11 3LW, UK
UKHW021957180926
475701UK00003B/71
* 9 7 8 0 3 6 7 8 6 5 8 2 5 *